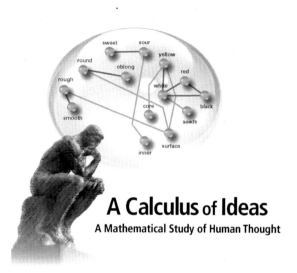

A Calculus of Ideas
A Mathematical Study of Human Thought

Immanuel Kant: "Human reason is by nature architectonic"

A Calculus of Ideas

A Mathematical Study of Human Thought

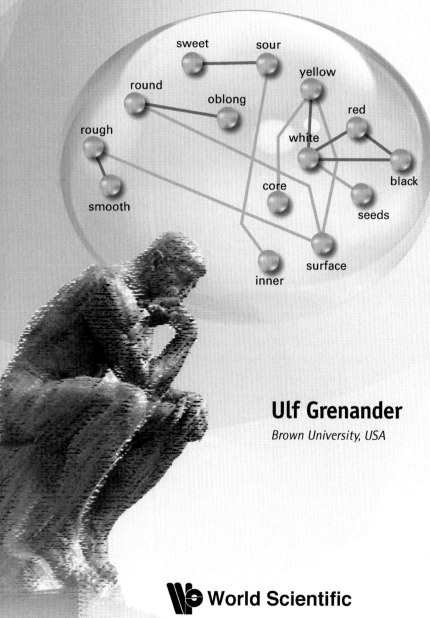

Ulf Grenander

Brown University, USA

World Scientific

NEW JERSEY · LONDON · SINGAPORE · BEIJING · SHANGHAI · HONG KONG · TAIPEI · CHENNAI

Published by

World Scientific Publishing Co. Pte. Ltd.

5 Toh Tuck Link, Singapore 596224

USA office: 27 Warren Street, Suite 401-402, Hackensack, NJ 07601

UK office: 57 Shelton Street, Covent Garden, London WC2H 9HE

British Library Cataloguing-in-Publication Data
A catalogue record for this book is available from the British Library.

A CALCULUS OF IDEAS
A Mathematical Study of Human Thought

ISBN 978-981-4383-18-9

Typeset by Stallion Press
Email: enquiries@stallionpress.com

Printed in Singapore

Foreword

The idea that the contents of the mind should be described as a vast graph has a long history. I think of it as first having been given a completely concrete form by the great English lexicographer Peter Mark Roget. He first compiled his graphical organization of all words of the English language "to supply my own deficiencies" in 1805. But I suppose the word got around that he had made an exciting compilation and he eventually allowed it to be published in 1852, calling it a Thesaurus. Especially in its earlier editions, Roget's Thesaurus describes the whole complex web of categories of objects, actions, social structure and mental states which we effortlessly access and play on in every single thought that comes into our minds. It has at its head a taxonomy reminiscent of Aristotle's categories but beneath these he creates a vast tree with tens of thousands of specific words with the lists which make explicit how they are linked to each other with multiple sorts of similarities. While philosophers speculated on ontological and epistemological issues endlessly and at great length, Roget did something no one else had dared to do: he looked at every word in the lexicon of English.

With the advent of Artificial Intelligence, a similar explicit list of all the elements of thought became essential. If we are to teach a computer to act and interact like a human, it must master this entire lexicon. More than that, an intelligent agent must master the connection of the lexicon to the world as revealed in the senses and motor interactions. For example, in Computer Vision, researchers today are asking *how many* categories of objects are there which their vision systems must be able to identify to match human visual competence — 1000? 10,000? Or 100,000? They started with trying to classify the 'Caltech 101' — a database of exemplars for merely 101 categories but are now hoping to master much greater numbers of categories.

Of course, the elements of thought do not live by themselves but are part of a very rich structure. Roget limited himself to mapping all the relations of similarity between pairs of words. Even these are surprisingly subtle. For example, there are many paths of length half a dozen or so from a word to its opposite, as Ron Hardin discovered. Here's such a path: generous $\leftrightarrow$ lofty $\leftrightarrow$ superior $\leftrightarrow$ exclusive $\leftrightarrow$ selfish $\leftrightarrow$ ungenerous. Then again, going back as far as early India when the famous Sanskrit grammarian Panini flourished, linguists have studied the links between sets of words which can be combined in a compound, a phrase or a sentence. Specifically case grammars as proposed by Charles Fillmore have been based on the fact that every verb has potential links with other words which embody specific objects and context for the action given by the verb — who acts, on what, when, where, etc. The resulting graph for any such grammar is what we call a 'parse tree'. A third type of link was investigated by early workers in AI. This is the link of inclusion between categories which they named the '*is-a*' link, as in 'a robin *is-a* bird', 'Socrates *is-a* man'. Their resulting graph was called a 'semantic net'.

Grenander, in this work on thought, has proposed and developed a general framework which incorporates all of these precedents. This framework expands on his extensive earlier work that he has named *Pattern Theory*. In this theory, the objects being studied are graphs whose vertices are called *generators* and which whose edges create *bonds*. In this book, the elements of thought are the *generators* and these are linked by bonds with values in a *modality lattice* analogous to the situation in a case grammar. A set of generators together with links provided by particular bonds make up the graph of a specific thought. This is not just an abstraction: the book below fleshes out these ideas with an astonishing variety of examples, drawn not only from everyday life but also (entertainingly) from history and literature. Grenander clearly loves to show how well his theory captures a huge variety of the thoughts that have come into his head.

In the history of theories of the mind, all too often people have treated our thoughts as though they consisted in rational logical calculations. Being made by scientists or philosophers, these theories sought to mirror what their makers felt they were doing in their heads — and they naturally assumed they were entirely rational and logical! It is only in the last few decades that economists have come around to the central fact that human behavior is often quite irrational and is not well modeled by goal of optimizing an objective function of prices. In a different direction, the necessity to model the analysis of noisy incomplete sensory data not by logic but by

Bayesian inference first came to the forefront in the robotics community with their use of Kalman filters. But this was only accepted in the computer vision community a decade after Grenander and his associates first introduced this approach to image segmentation.

In this book, Grenander comes to grip with both of these key insights. On the one hand, a central part of this theory is the introduction of probability measures which describe what associations and deductions we are likely to make. And on the other hand, he studies our emotions and incorporates these centrally in his model of our thoughts.

This book is a "proof of concept". As Grenander is the first to admit, the project of formulating a definitive set of generators and the modalities of their bonds is a huge and challenging project. But he has laid out the path and thrown down the gauntlet challenging the next generation to fully realize his extraordinary vision. I, for one, agree with Ulf that this is perhaps the most important task on the road to understanding our minds today.

David Mumford
2012

Preface

This monograph reports a thought experiment with a mathematical structure intended to illustrate the workings of a mind. It is speculative rather than empirical, based mainly on introspection, so that it differs radically in attitude from the conventional wisdom of current cognitive science. No doubt this will cause a negative reaction from many readers. My only defence is that the elegant simplicity of the proposed structure will make it seem plausible, indeed likely to be true, as a representation of high level thought processes. It would be presumptive to claim that it will eventually be accepted as the correct model, but I hope that the very attitude will turn out to be productive.

I am grateful to my colleague Yuri Tarnopolsky for many stimulating discussions and for his thoughtful comments and criticism. He had also helped with the preparation of the manuscript. He played an important role in developing this theory. Sahar Primoradian helped in extending and improving the software by imaginative coding. My old friend Chii-Ruey Hwang made it possible to have it published.

This manuscript has had a long gestation, starting in 2006, during which my wife gave me wonderful moral and physical support — without it, this study could not have been completed. I also thank my editor Lai Fun Kwong for her careful work.

Ulf Grenander
2012

Abstract

Can the human mind be understood? And understood in what sense? In the natural sciences we understand a phenomenon when we have reduced it one level into concepts that seem more elementary. For example, from the high level of continuum mechanics we descend one level to atomistic structures, from atomism one level down to electrons, protons, then to the quark level?

We shall reduce mind activities to elementary ideas and mental operations on them. Connecting the ideas according to certain rules we will be led to an algebraic structure, one for each cultural sphere. Within such a sphere individuals are characterized by the usage they make of the mental concepts as expressed by a probability measure resulting in a calculus of ideas.

To illustrate this somewhat abstract reasoning we shall build some artificial minds constructed according to the algebraic rules and then superimpose adequate probabilities on them. The minds will be realized by software: running the software the resulting thoughts can be displayed so that the user gets an impression of a personality. With thoughts we mean not just logical reasoning but above all emotions and body sensations.

We shall not study the physical substrate for thinking, the neural system, but the very fact that it has the form of a network implies that thought also has graph structure. This in turn suggests that language and its grammars, evolved later, also should be expressed in terms of graphs.

All of this is highly speculative with little empirical support except for the important, and not always appreciated, tool of introspection. We shall venture into even more speculative reflections on possible neural structures that would be consistent with our calculus.

Contents

Part I
An Architecture for the Mind

Chapter 1

Introduction

1.1 A Mathematical Theory of Mind?

The human mind is a mystery. Although it is so close to us — we live in and with it — we do not really understand how it works. Philosophers and thinkers in general have struggled with this question for millennia and much has been learned, most in a vague and unspecific form. Some attempts have also been tried to describe it through logical schemata and in mathematical form. But human thought is (normally) not completely rigid; it is only partly predictable.

We instinctively avoid believing that our thoughts are generated by a more or less mechanical device. We do not want to be seen as machines. Hence we tend to reject statements like the one by Karl Vogt, a 19th century German philosopher, who stated that "the brain produces thoughts as the liver produces bile, or the kidneys produce urine". But few would deny that the material substrate of thought, the neural system of the brain, obeys the laws of physics/chemistry, so that it is not impossible that *there may exist mathematical laws of thought* in principle derivable from physic/chemistry. Such laws would have to be probabilistic. The following consists of speculations with no firm support in empirics, just ideas that seem plausible (to the author).

We shall consider thought processes that include logical thinking, but this is only one mode among many. We follow Damasio (1999) who discusses the dominating role of emotions for human thought in an elegant and convincing way. We shall include fear, love, emotions... But recall Pascal's dictum: "The heart has its reasons, of which reason knows nothing." Indeed, we know only little about the functioning of emotional thought processes. But wait! *We are not after a general theory of human thought,* indeed we do not believe in such an endeavor. Instead we will try to present only a shell, a scheme only, of human thought that will have to be filled with content different for each individual, setting different values to the (many) mind parameters. This content can have its origin in the genetic and cultural background in which the individual lives, as well as being formed by experiences leading to a dynamically changing mind. Thus we will concentrate on the *general architecture* of the building rather than on its detailed specification of mortar and bricks.

We shall deal with the mind without reference to the brain. A completely reductionist mind theory would be based on neuro-physiological knowledge, deriving mental processes from what is known about their cerebral substrate. We are certainly in favor of such an approach, but in the absence of a complete brain theory, it is not feasible at present. Instead we shall base the construction on introspection and on what has been learned over the centuries in a less formal setting about the working of the mind by clinicians and what can be found in novels, poetry and plays. This non-positivist attitude is open to the criticism that it leads to no testable hypothesis. We admit that this is true, at least in the immediate future, and accept the criticism.

The last several decades have witnessed remarkable process in the neurophysiology of the brain — many elegant experiments have thrown light on the functioning of neurons, at first for single neurons and more recently for cell assemblies. This has led to an impressive body of empirical knowledge about the brain. Some researchers have tried to increase our understanding of the human mind through mathematical studies of the firing rates of neurons. It seems doubtful to this author whether mathematical work of this type *alone* will lead to more insight in the human mind than what the purely experimental results have shown. This author is all in favor of such a reductionist approach: it is necessary — but not sufficient! Perhaps such studies can help in understanding how *Ratus ratus* runs in mazes or how we turn our right hand at some command, but for the understanding of the mind of *Homo sapiens* they are flagrantly insufficient. We are aware of the many talented and knowledgeable researchers applying mathematical analysis to neural rates, concentrating on neural behavior while neglecting high level activities of the human mind. They seem suspicious of a theory of higher mental faculties. Alas, they include even such personalities as *sagax* Mumford. We beg the indulgence of those researchers, if we put more trust in the introspective wisdom of Aristotle, Shakespeare and William James (perhaps also that of his brother), as well as in the collected clinical experience of psychiatrists/neurologists, when it comes to describing and analyzing the high level mental activities. Expressed differently, our approach could perhaps be stated as studying the software of the mind rather than the hardware.

1.2 Substance and Change

We will base our construction of PoT (Patterns of Thought) on the principles of General Pattern Theory, GPT.[1] To help the reader we will now offer a brief and clearly superficial introduction to the basic ideas in GPT. Let us start from the proposition

$$General Pattern Theory = Substance \oplus Change. \tag{1.1}$$

1.2.1 *Substance*

The Substance of GPT, the generators, are the building blocks that will be transformed and combined to form regular structures. For concreteness, we shall give a number of examples, starting with some simple ones, that will reappear later.

1. Triangles and other elements of Euclidean geometry.
2. Abstract symbols: letters, names and concepts.
3. Audio sequences: music or speech.
4. Picture elements: B/W or colored 2D pieces with curved boundaries.
5. Concepts: animate objects, activities, properties, ...
6. Rewriting rules as they appear in context free grammars.

But the generators are not just such objects as the above. In order to be subject to change they must also be equipped with other properties, to wit bonds regulating how the generators can be combined together. We shall denote the set of generators used in a particular situation by G and the generators generically by $g, g_1, g_2, \ldots$.

---

[1]See Grenander (1993) in References.

1.2.2 *Change*

First we shall let the generators be changed by operations: elements of a similarity group. Here are some specific examples.

1. Consider triangles in the plane and change them by translations. The similarity group is $S = TRANS(2)$. It will look something like Figure 1.1.

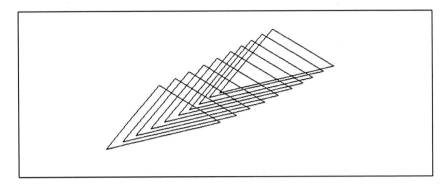

Figure 1.1: Translation group

Let us also allow rotations as similarities, $S = SE(2)$, the Special Euclidean Group in the plane. (Special means that reflections are not allowed.) It can look like Figure 1.2.

Now we also add uniform scaling as shown in Figure 1.3.

2. For abstract symbols we shall choose the similarities as a permutation group. For example the symmetric (permutation) group over 7 objects in Figure 1.4.

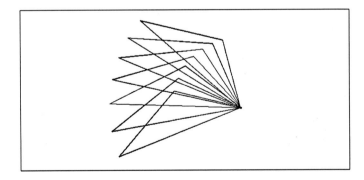

Figure 1.2: Rotation group

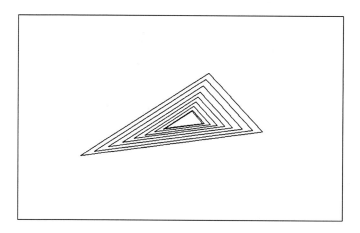

Figure 1.3: Scaling group

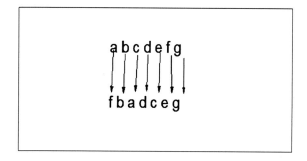

Figure 1.4: Symmetric group

3. If the generator is a music file we could select S as addition mod(12), moving semi-notes up or down. In other words as musical transposition: from F major to D major; see Figure 1.5.

4. Operate on a set of pieces by trying to combine them so that they fit both in shape and coloring. Thus we are dealing with a jigsaw puzzle and we try to solve it. See Figure 1.6.

This example illustrates well how *Pattern Theory starts from simple "pieces" and then combines them together so that they fit.* Bond values are here the boundary of the pieces together with the corresponding values along these contours.

5. We can deal with concepts more or less in the same way as for abstract symbols; see above.

Figure 1.5:

Figure 1.6:

6. For CF grammars we can combine the generators into a graph (here a tree). Starting from a simple context-free grammar, we derive a sentence by connecting the rewriting rule in the context-free grammar, see Figures 1.7–1.11.

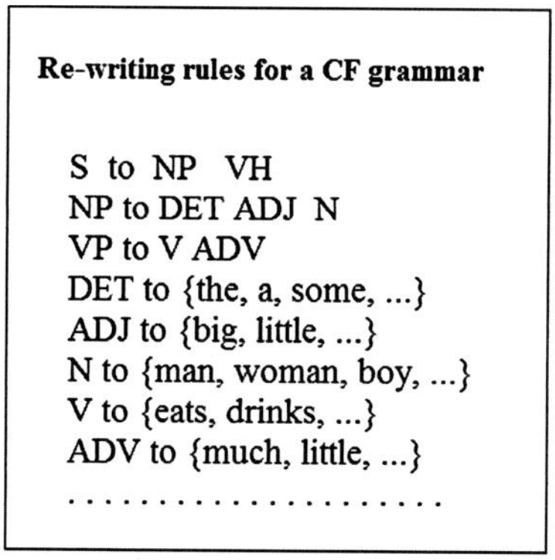

Re-writing rules for a CF grammar
S to NP VH
NP to DET ADJ N
VP to V ADV
DET to {the, a, some, ...}
ADJ to {big, little, ...}
N to {man, woman, boy, ...}
V to {eats, drinks, ...}
ADV to {much, little, ...}
. . . .

Figure 1.7:

Start derivation of sentence
S

Figure 1.8:

S
NP
VP

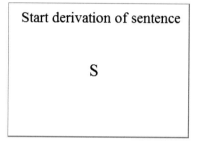

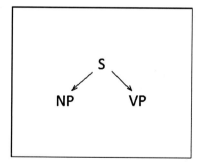

Figure 1.9:

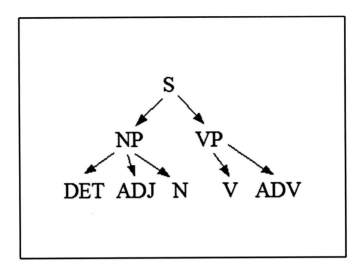

Figure 1.10:

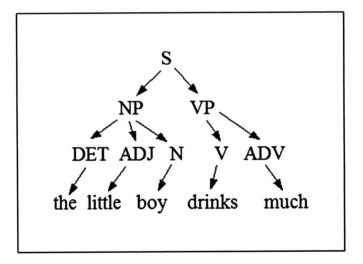

Figure 1.11:

An observant reader will have noticed that *Change* here means two different things. For cases 1, 2 and 3 we operate on individual generators with elements of the similarity group S. On the other hand, for cases 4, 5 and 6 we combine several generators into configurations, in which generators

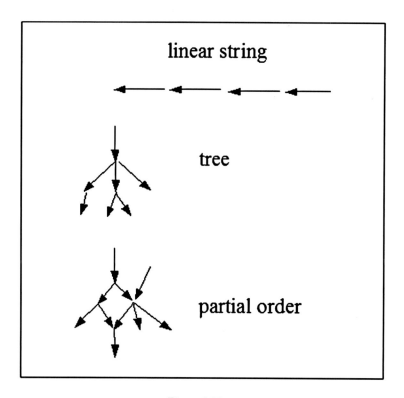

Figure 1.12:

are connected to each other according to the topology of a graph that will generically be denoted by *connector* = σ. The σ's shall be selected from a *connection type* Σ, $\sigma \in \Sigma$. Symbolically:

$$Change = Similarities \oplus Combinations. \qquad (1.2)$$

Three such connection types are illustrated in Figure 1.12.

But configurations will not be arbitrary collections of graphs (*from* Σ) of generators. Instead we shall only use *regular* configurations meaning that if two generators are connected in σ they must fit according to some bond relation ρ. In other words, if a bond β_1 of a generator g_1 is connected to a bond β_2 of g_2 then we must have $\rho(\beta_1, \beta_2) = TRUE$. In the construction of MIND in the following the bond relation will be expressed via a modality transfer function.

Hence we are dealing with local regularity, ρ, global regularity, Σ, and total regularity, $\mathcal{R} = < \rho, \Sigma >$.

1.2.3 *Patterns*

In everyday language we use the term pattern in a vague way but now we shall give it a precise meaning. We shall say that a set $\mathcal{P}$ of configurations forms a pattern if it is invariant with respect to the similarity group used: $s\mathcal{P} = \mathcal{P}; \forall s \in S$. First a simple example: All right angled triangles in the plane, $S = SE(2)$.

Then a more complicated example:

Generator space $G = HUMANM \cup HUMANF$
$$\cup \ CHANGEHANDS \cup OBJECT$$

with

$$HUMANM = Bob, Dick, Charles, \ldots$$
$$HUMANF = Mary, Ann, Carin, \ldots$$
$$CHANGEHANDS = give(3), take(3), borrow(3), \ldots$$
$$OBJECT = book, flower, box, \ldots$$

giving rise to the configuration (and many others).

Now form the *set* of configurations $\mathcal{P}$ with a similarity group S that permutes ideas belonging to the same modality.

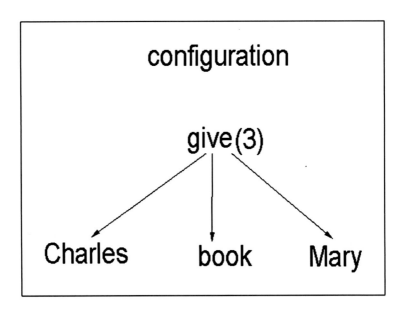

Figure 1.13:

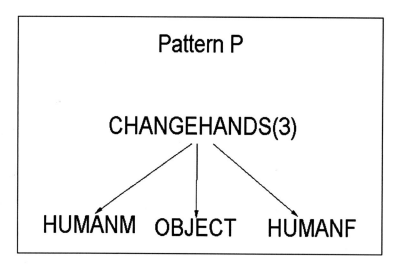

Figure 1.14:

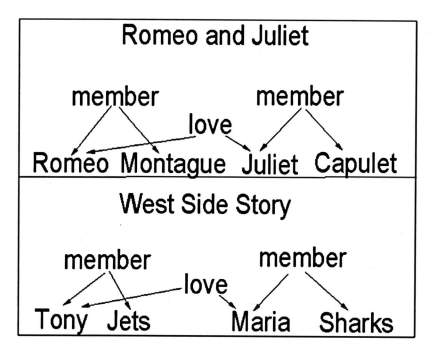

Figure 1.15:

This sort of construction will be used often in the following.

If we choose a similarity group that permutes names, then the two configurations in Figure 1.13 are S-equivalent: they belong to the same pattern.

A pattern can look like Figure 1.14 with abstract classes of concepts rather than concrete ones like 'give(3)'.

Now we are ready to apply the concepts of Pattern Theory to human thinking. A well-known pattern is seen in Figure 1.15.

Chapter 2

Creating Thoughts: Algebra of Thinking

2.1 What We Shall Do

Our goal is to build a model of the mind in pattern theoretic terms: Starting from simple, atomic, mental entities (the generators of pattern theory) we shall combine them into regular structures, thoughts, (configurations) later on to be controlled by probabilistic rules of connections. In this way patterns of thought will be built *pace* Kant as hierarchies of more and more complex structures for which we shall introduce a *calculus of ideas*. Note that we are aiming for representations of thoughts of different types: deductive reasoning (including logical errors), feelings like love and hate, doubts and questions and many others. We will be guided by David Hume's radical proposition:

"Though our thought seems to possess this unbounded liberty, we shall find, upon a nearer examination, that it is really confined within very narrow limits, and that all this creative power of the mind amounts to no more than the faculty of compounding, transposing, augmenting, or diminishing the materials afforded us by the senses and experience."

A statement that is still valid. By senses we mean not only the classical five: vision, audition, smell, touch and taste, but also sensations due to hormonal and other body attributes such as affects, feelings, hunger, muscular activity, etc., following Damasio (1999). And, of course, some thinking takes the form of pictures. Hence ideas are not necessarily represented by words and sentences in a natural language, so that our approach is *extra-linguistic*. Thinking comes before language!

Thoughts could be, for example,

"John loves Mary"

"smell of Madeleine cake"

"index finger hurts"

"bell tolls"

We shall limit ourselves in this book to outlining a mathematical representation theory but hope that sooner or later it will be applied to knowledge available to experimental neurologists/psychiatrists.

So we shall search for answers to the following questions:

What are the mental objects that make up the mind?

What are the mental operations that act upon these objects?

How do these objects combine to form thoughts?

2.2 Judging a Mind Model

> Carver Mead: "... *you understand something*
> *when you can build it*"

But here is the rub. Since we are admitting that our mind model does not rely on firmly established facts, neither on neurophysiological theory, nor on objective cognitive facts, how are we going to judge it? What criterion will be applied to evaluate its validity? It is easy and tempting to speculate, but without self criticism we will have no guarantee that we have achieved more than an amusing (?) thought experiment. It is tempting to get immersed in abstract and too general speculations: here, as elsewhere, the devil is in the details. But we shall spend much time on working out the details.

Appealing to Carver Mead's motto we shall *build* a mind model in software, expressing our theoretical constructs in program modules. We shall be satisfied with the model, at least temporarily, if the program executes in a way that seems reasonably close to what our intuition expects of a particular human mind. This is somewhat related to Turing's celebrated test, but our goal is less ambitious. We are not in the business of artificial intelligence, we do not intend to create intelligence or a simile of it. Instead, our more modest goal is to present a shell that can be filled with specifically chosen entities resulting in a coherent scheme consistent with what we believe is human thought.

In passing we mention Joseph Weizenbaum's celebrated program ELIZA that mimics conversation between a patient and an analyst. It attracted a lot of attention, even a belief in the psychotherapist it simulates, to the extent that its inventor came to be surprised and even embarrassed by the misguided enthusiasm that the ELIZA program generated. The code supporting the program is simple, but the behavior is, at first, quite impressive. What we are after, however, is a code that rests on a pattern theoretic analysis of the human mind specifying the details of mental processes, and such codes will have to be complex.

As we shall see it will take some complex software to achieve our goal, even roughly. To facilitate programming we shall write in MATLAB although this will result in slow execution. In a later stage we may compile the code into C++ or into executables, but at the moment we are not concerned with computational speed.

2.3 Mental Architecture

Hence we shall *build* mind states from primitives, elements that express simple mental entities: feelings and emotions, thoughts about the external world as well as about the inner self, doubts and assertions, logical deductions and inferences. We shall allow the reasoning of the mind to be incomplete, inconsistent and, well, unreasonable. Influenced by Damasio (1999), and perhaps by Vygotskij (1962), we shall include feelings, perhaps originating outside the brain, and their interaction with conscious thought. We shall be guided by introspection, our own of course, but also by that of others accumulated over eons in novels, poetry, plays. Perhaps we can also find help in figurative paintings and other art forms. In addition, a multitude of philosophers and psychologists have offered insight into the working of the human psyche in a more technical sense. Recently, scholars in cognitive science and artificial intelligence have presented schemes for the understanding of natural and man-made minds, often in a controversial form. We shall borrow from many of these sources, sometimes without explicit attribution. The basic idea in what we shall be doing, however, was suggested in Grenander (1981).

There is a huge literature on modeling the human mind. Here we shall just refer the reader to Appendix A for a sketch of a few of the attempts in this direction.

ADVICE TO THE READER: The next section is more abstract than the rest of the book. Therefore, the reader should just skim it and perhaps return to it later. It will be illustrated more concretely later.

2.4   An Algebra of Human Thought

> Wittgenstein: "*The picture is a model of reality.*
> *To the objects correspond in the picture*
> *the elements of the picture.*
> *The picture consists in the fact that its elements are*
> *combined with one another in a definite way.*"

Let us begin with an axiomatic description of the algebra, to be followed by a concrete discussion elucidating the axioms and introducing the concepts needed for the following.

2.4.1   *Primitive Ideas*

Thoughts are formed as compositions of generators (primitive ideas), in some generator space, $g \in G$. G is finite but its cardinality can vary with time as the mind develops. A generator g has an arbitrary (variable) number of in-bonds with the same bond value $\beta_{in}(g)$, and a fixed number $\omega_{out}(g)$ of out-bonds with bond values $\beta_j(g); j = 1, 2, \ldots, \omega(g)$.

2.4.2   *Modalities*

Bond values are from a lattice $\mathcal{M}$ of subsets of G.

2.4.3   *Similarities of Ideas*

On the generator space G there is defined a permutation group S, the modality group. Two generators g_1 and g_2 are said to be similar if $\exists s \in S \ni g_1 = sg_2$. The s-operation preserves bonds.

2.4.4   *Compositions of Primitive Ideas*

A thought is a labeled acyclic directed graph thought $= \sigma(g_1, g_2, \ldots, g_n); g_i \in G$ where the connector graph σ connects some jth out-bond $\beta_j(g_{i_1})$ of generator g_{i_1} to an in-bond of generator g_{i_2}. The modality group is extended to thoughts by: s thought $= \sigma(sg_1, sg_2, \ldots, sg_n)$.

2.4.5   *Regular Thoughts*

A thought is said to be regular if only out-bonds connect to in-bonds carrying the same bond value: regularity $\mathcal{R}$. The set of all regular thoughts for specified $G, \mathcal{M}, \ldots$ is called $MIND(\mathcal{R})$. *A given set* $\{MIND(\mathcal{R}), P\}$ *is called a personality, where P is a probability measure on* $MIND(\mathcal{R})$.

2.4.6 Thought Patterns

A subset $\mathcal{P} \subset MIND(\mathcal{R})$ is called a thought pattern if it is invariant with respect to the modality group S.

2.4.7 Completion

Thoughts are made meaningful by the application of the COMPLETE operation that closes out-bonds.

2.4.8 Generalization

Thoughts are generalized by the application of the MOD operation from a semi-group GENERALIZATION.

2.4.9 Abstraction

The device of encapsulation abstracts thoughts to ideas that can be referred to as independent units; they are automatically added to the generator space G.

Also we shall appeal to a

PRINCIPLE OF ISOLATION: *The MIND strives to make thoughts meaningful so that they can be standing alone; hence they should be complete (see below for this concept). We can speak of a completion pressure.* The environment contains things, but also events that are happening or have happened, and other non-physical facts. Recall Wittgenstein's dictum: "the world consists of facts, not of things", *Tractatus Logicus-Philosophicus* (see References). We shall include physical things like

$\{dog,\ cat,\ human,\ John,\ table,\ car, \ldots\} \subset G$

but also non-physical ideas like

$\{thought,\ hate,\ walk,\ fear,\ say, \ldots\} \subset G$

and events like

$\{wedding,\ fight,\ transaction\} \subset G$

to mention but a few.

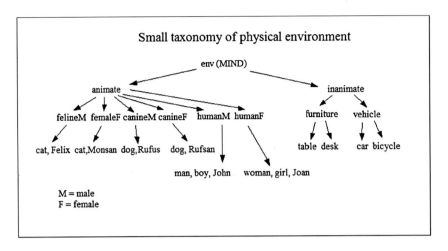

Figure 2.1:

But how should we organize such generators? One way is to order them through a Linnean taxonomy in organizational trees like the one shown in Figure 2.1 (or forests). In it we have shown the physical environment, $env(MIND)$, at the root (top of figure) of a tree. Paths are like $env(MIND) \rightarrow animate \rightarrow felineM \rightarrow cat \rightarrow Felix$.

Most of the elements in this taxonomy are self-explanatory, with one exception: note that the generator "canineM" is a generic symbol for male dogs in general, while "Rufus" signifies a particular dog. The observant reader will notice, however, that in order that this represent a real partition, the set "canineM" must be defined as different from "Rufus". We shall return to this later.

Non-physical generators are at least as important as things. For example, $g = think$ representing someone's thinking, or $g = say$ meaning a statement is being made by someone. Here that *someone* can be "self" or another human member of G. There will be many non-physical generators: "doubt", "question", "answer", "write", and so on. Combining them we get diagrams like those in Figure 2.2 where the interpretation of a diagram is given on the right side. We have used notation "think1" to indicate that it has one arrow (out-bond) emanating out from it, "question2" has two arrows from it and so on, so that "question2" is different from "question3". This is formalized through the notion of arity to be discussed in Section 4.2.

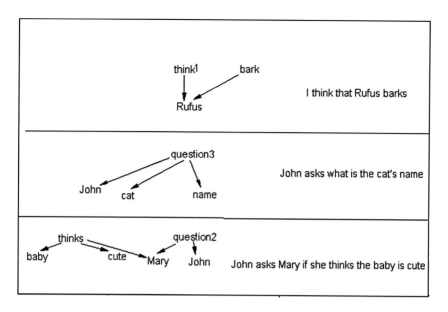

Figure 2.2:

2.4.10   *Caveat*

It is tempting to think of the generators as words and the diagrams as sentences, but this is not at all what we have in mind. Recall the famous Sapir-Whorf hypothesis: "... the fact of the matter is that the *real world* is to a large extent unconsciously built up on the language habits of the group" and that our thought processes are directly or indirectly made up of words. We do *not* subscribe to this hypothesis. On the contrary, our construction of a formal mind will be done independently of language to the extent that this is possible. It is not easy to free oneself from the straitjacket of language, but we shall try to do this in the following to the extent it is possible. We shall deal with *concepts* — *not words*. Actually, we will be forced to use notation more precise than words alone. As an example we may distinguish between generators like $g_1 = activity_1$ and $g_2 = activity_2$, with the usage of g_1: "John works" and of g_2: "John works with a hammer"; see the remarks at the end of last section. We shall make many such distinctions and insist that they are more than mere technicalities; they are needed in order that the mind representation be precise. But we do not insist that the mind and its thinking be precise, it is not, only that our representations of the thinking be precise. In spite of the conventional wisdom, we proclaim

CONCLUSION: *Thinking comes before language, it is the primary mental activity.*

To exemplify the above: the meaning of the generator $g = dog$ is reasonably clear, while $g = question$ requires some explanation. It is certainly not the word "question" itself; instead we intend it to represent the *act* of questioning, someone asks someone else about something; the notation "question3" would be more adequate.

Therefore we shall strive for a *language independent* mind theory, admitting that we have only partially realized this goal, an extra-linguistic representation of a mind.

2.4.11   *Levels, Modalities, and Arities in Mind Space*

In Figure 2.1 we have arranged the generators in *levels*: $g = catM$ is situated below $g = felineM$ which is on the next higher level in the taxonomy partition. But we shall formalize the concept of *level* in another way. We first introduce the concept of *modality*.

The generator space will be partitioned into a family $\mathcal{M}$ of subsets, modalities $M(m) \subset G; m = 1, 2, \ldots, card(\mathcal{M})$,

$$G = \cup_{m=1}^{card(\mathcal{M})} M(m) \tag{2.1}$$

together with a *partial* ordering so that $m_1 \downarrow m_2$ for some pairs $m_1, m_2 = 1, 2, \ldots, M$ while other pairs may not be ordered with respect to each other. A modality will contain generators with related meaning, for example

$$color = \{red, \ blue, \ green, \ yellow, \ldots\} \in \mathcal{M} \tag{2.2}$$

or

$$movement = \{run, \ jump, \ turn, \ still, \ldots\} \in \mathcal{M} \tag{2.3}$$

where the set of all modalities has been denoted by $\mathcal{M}$ and enumerated $m = 1, 2, \ldots, card(\mathcal{M})$. This is the *modality lattice*. Occasionally we shall make use of the TERM *modality mixes*, meaning unions of modalities. An example of a modality mix is $action1 \cup action2$. An extreme modality is $m = mod = \mathcal{M}$ itself, all modalities together. Modalities are denoted by capital letters in contrast to the primitive ideas which are written with lower case letters.

The generators $g_1 = bark$ and $g_2 = dog$ are naturally ordered, $g_1 \downarrow g_2$, but $g_3 = yellow$ and $g_4 = smooth$ do not appear to have any natural order. Thus the ordering is partial rather than completely.

With the convention that all 'object'-generators, animate or inanimate, are put on *level* one we shall use the

DEFINITION: *The level level(g) of a generator g is the shortest length l of regular chains*

$$g \downarrow g_{l-1} \downarrow g_{l-2} \downarrow g_{l-3} \downarrow \ldots \downarrow g_1; level(g_1) = 1. \qquad (2.4)$$

Thus a generator g with $l = level(g) > 1$ can be *connected downwards* to a number of generators on level $l-1$. We shall need a concept characterizing the connectivity of generators, namely the *out-arity*, sometimes to be called down-arity.

Behind this construction is the PRINCIPLE OF ISOLATION. The primitive (elementary) thoughts on level 1 can stand alone and still be meaningful. The concept of *new idea*, to be introduced later, is meant to be meaningful standing alone; hence it should belong to level 1. For a primitive thought to be on level L it should be possible to make it meaningful standing alone by adding primitive thoughts from level $L - 1$ and lower.

DEFINITION: *The number of generators that can be connected directly downwards from g is called the arity $\omega(g)$ of g.*

In particular the generators on level 1, the 'things', all have arity 0. Hence $g_1 = bark$ in Figure 2.2 belongs to level 2 and arity 1, while $g_2 = Rufus$ belongs to level 1 and arity 0. But we need more information about the connectivity of generators. If $\omega = \omega(g) > 0$ we must decide to what other generators it can connect. This is the purpose of *bonds*, more precisely downward bonds. To each generator g and its downward jth bond we associate a subset of G denoted $\beta_j(g) \subset G; g \in G; j = 1, 2, \ldots, \omega(g)$. We shall choose the subsets as modalities or modality mixes. For example, we may choose $\beta_1(love) = humanM$ and $\beta_2(love) = humanF$ for a heterosexual relation. The up-bonds will be the modality of the generator itself.

Of special importance are the "regular modalities", i.e. modalities such that its generators have the same arity and level that will lead to regular thoughts. The others, the irregular modalities, will be used for taxonomy but not for the formation of meaningful thoughts. In Appendix C the regular modalities are shown as rectangular boxes, while the irregular ones are indicated as diamond shaped boxes.

Modalities can be ordered by inclusion. For example, $ANIMAL \subset ANIMATE$. Note that this ordering is different from the partial order discussed above. It is clear that $\mathcal{M}$ forms a lattice, a POSET. This means that the ordering of modalities produces entities on different *planes of modality*. We have been denoting modalities (on the first plane) by capital letters and shall use bold faced capitals for the next plane. For example, we could have:

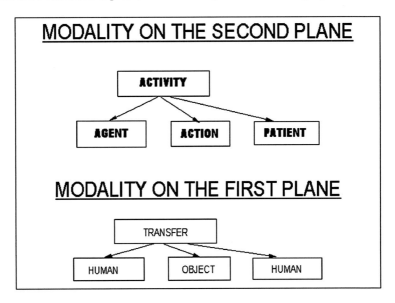

Figure 2.3:

2.4.12   *A Concept of Concept*

We shall make the idea of a modality clearer. A concept, a modality M, is an item that can be used as an independent unit: it can connect to primitive thoughts as well as to other modalities as long as regularity is observed. The size of the set $M \in G$ can be just 1, but it should be bigger in order to serve as a concept useful for abstract thinking. As an example look at Figure 2.4.

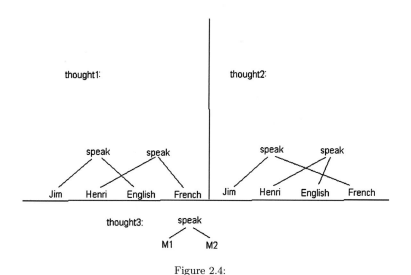

Figure 2.4:

In this figure, *thought*1 means that Jim speaks English and Henri speaks French, while *thought*2 says that Jim speaks French and Henri English. If *thought*1, *thought*2 $\in$ *MIND*, we could form the modality $M1 = \{Jim,\ Henri\}$ and $M2 = \{English,\ French\}$ and consider *thought*3 regular, *thought*3 $\in$ *MIND*. But if *thought*1 $\in$ *MIND*, *thought*2 $\notin$ *MIND*, the creation of the modalities $M1$, $M2$ would not be legal. We would have to introduce the contrived primitive ideas $speak^1$ and $speak^2$, the first one with out-bonds (Jim, English) and the second one with (Henri, French).

2.4.13 *Regularity of Mind States: Conformation of Thoughts*

> H. Poincare: "... *ideas hooked together as links in a chain* ..."

Now let us combine generators (elementary ideas) into *configurations*, or thoughts, represented by diagrams like those in Figure 2.2 and written formally as

$$thought = \sigma(g_1, g_2, \ldots, g_i, \ldots, g_n) \tag{2.5}$$

25

where σ is a graph joining the generators $g_1, g_2, \ldots, g_n$ in the diagram. In the first configuration in Figure 2.2 the diagram has three *sites* called "think", "Rufus" and "bark", meaning "I think that Rufus barks". This graph has two *edges*, namely $1 \to 2$ and $1 \to 3$. We shall use subscripts $i = 1, \ldots, n$ and so on to enumerate the generators, and $j = 1, \ldots, m$ and so on for the edges (also called connections) of the graph (for the down-bonds of a single generator). Hence, with the notation

$$n = size(c), \quad m = size(\sigma) \tag{2.6}$$

we have $n = 3, m = 2$.

A central concept in Pattern Theory is that of *regularity*. In the following we shall use two types of regularity:

DEFINITION: *A configuration thought* $= \sigma(g_1, g_2, \ldots, g_i, \ldots, g_n)$ *is said to be COMPLETELY REGULAR if any jth down-bond $\beta_j(g_i)$ of any generator g_i in it is connected to a generator g_i' satisfying the bond relation*

$$\rho : g_i' \in \beta_j(g_i) \tag{2.7}$$

and a weaker concept:

DEFINITION: *A configuration, or thought, $c = \sigma(g_1, g_2, \ldots, g_i, \ldots, g_n)$ is said to be REGULAR if any connected jth down-bond $\beta_j(g_i)$ satisfies the bond relation*

$$\rho : g_i' \in \beta_j(g_i). \tag{2.8}$$

In other words, a completely regular configuration has all its down-bonds connected, but an incomplete has some down-bond left open. In Figure 2.2 the second configuration is complete but if the connection *question* $\downarrow$ *cat* is broken it is incomplete (assuming that $\omega(question) = 2$).

We shall use the terms *complete* and *incomplete thoughts* when talking about configurations. When the configuration is made up of a single generator g it is called a *primitive (or elementary) idea*.

An incomplete or irregular thought may not have any acceptable interpretation and will therefore not always reach the level of consciousness. Nevertheless we shall study them, in accordance with our goal of studying thinking in all its imperfections, lack of consistency and with logical mistakes. At any instance there is a *chatter of competing thoughts* most of which will not reach the conscious level. More precisely, an incomplete thought, an irregular configuration of elementary ideas, will have a high energy (low probability; this will be explained later). It will therefore quickly be deleted or modified to lower the energy; if it appears at all in consciousness it would be only for a fleeting moment. Later on we shall show such *chattering of incomplete thoughts* in the configuration diagrams.

The set of all regular configurations is called the (regular or completely regular) *configuration space*, the MIND, and is denoted by $MIND(\mathcal{C}(\mathcal{R}))$; it represents the set of all the thoughts that this mind is capable of. Note that the *regularity requirement of an idea means that its constituent sub-thoughts (ideas) conform*. Hence we view thoughts as geometric constructs, to wit graphs, whose composition expresses the personality of the individual. Following the wise Spinoza (1670), (*"ordine geometrico"*), we claim the

CONCLUSION: *Human thought has geometric structure.*

Note also the resemblance to chemistry. This has been observed and elaborated by Tarnopolsky (2003). Generators correspond to atoms, configurations (thoughts) to molecules, and bonds to bonds. Sometimes we shall use chemistry like notation as $idea^l_{omega}$. We shall distinguish between different *isotopes of ideas*, for example $give^2_3$ and g^2_2. This notation will be used only when clarity demands it.

CONCLUSION: *The structure of the MIND is hierarchic, architectonic, with composition as the fundamental operation.*

2.4.14 *Creation of New Ideas*

The MIND will be dynamic in that the generator space is not static, it changes over time. A complete thought (recall: no unconnected out-bonds) can be made into an independent unit, a new generator that can be dealt with without reference to its internal structure. Hence *thought* $= \sigma(g_1, g_2, \ldots, g_n)$ can be made into an *idea*, a new generator added to G on level 1 and hence with no out-bonds. We can think of this procedure as an *encapsulation process.*

For example, the complete thought in Figure 2.4a means that one should love one's neighbor. When encapsulated it becomes a new generator that could perhaps be named "CommandX", but in the automated working of the mind we shall use more neutral notation like $idea_k \in G$ with a counter k.

Now let us make this more precise. Say that the MIND has produced a conscious *thought* with the size $n = size(thought)$, and the generators $g_1, g_2, \ldots, g_n$. With some probability $p_{create}(n)$ we shall abstract *thought*

Encapsulation operation to create new idea in G

Figure 2.4a:

to a new idea $idea_k \in G$, where k is a counter that will be successively updated as new ideas are created. The probability distribution $\{p_{create}(\cdot)\}$ expresses the sophistication of MIND: if it allows big values of n with considerable probabilities, the MIND is capable of powerful abstraction and vice versa.

If the MIND's decision is "create", a new idea is created and it will be put in a new modality *COMPLEX* on level 1, since it can stand alone, with the in-bond *idea*. The observant reader will have noticed that this differs slightly from our convention for defining modalities but will be useful for the coding.

Of course new ideas can also be created from sensory inputs, but the most interesting ones occur higher in the hierarchy when abstract concepts are created.

CONCLUSION: *MIND creates new ideas using the operation ENCAP-SULATION.*

2.4.15 *Patterns of Thought*

Following the general principles of Pattern Theory[1] we introduce a *similarity group S*, the *modality group*, on the generator space G:

$$S = S_1 \times S_2 \times \cdots \times S_m \times \cdots \tag{2.9}$$

where S_m is the permutation group, the symmetric group, over the set of generators in the regular modality $m = mod \in \mathcal{M}$. If two generators g_1 and g_2 are *similar* in the sense that there is a group element $s \in S$ such that

---

[1]See GPT, Chapter 1.

$g_1 = sg_2$, it is clear that this similarity induces a partition of the generator space into modalities as equivalence classes.

For example, $g_1 = $ "*John*" and $g_2 = $ "*Jim*" may be equivalent but probably not $g_1 = $ "*John*" and $g_2 = $ "*Mary*": this is an expression of the principle "arbitrariness of the sign" to quote de Saussure although he spoke of language rather than thought. This *modality group* enables the mind to *substitute mental entities* for another, i.e. abstract thinking, but preserving modalities, and avoiding incorrect references by not allowing primitive idea to be substituted for more than one other primitive idea. Hence the substitutions do indeed form a bijective map: a permutation within modalities.

As in all algebras, homomorphisms play an important role the calculus of thought.[2] The above transformations constitute configuration homomorphisms.

Also form subgroups of S *over the modalities* $m_1, m_2, \ldots$

$$S_{m_1, m_2, \ldots} = S_{m_1} \times S_{m_2} \times \cdots . \qquad (2.10)$$

A set T of thoughts, $T \subset MIND$ is called a *thought pattern* if it is invariant with respect to the modality group S. It is called a (*restricted*) *thought pattern over the modalities* $m_1, m_2, \ldots$ if it is invariant with respect to the similarities over these modalities. Thus all modalities are thought patterns but we shall encounter much more complicated patterns in what follow. Two examples are shown in Figure 2.5.

The set of all thought patterns in MIND will be denoted $\mathcal{P}$. It represents the power of MIND's ability of abstract thinking.

In General Pattern Theory a clear distinction is made between configurations and images.[3] While a configuration specifies generators and connections between them, an image is what can be observed. This is analogous to the distinction between a formula and a function in mathematics. For the elements in the MIND the identification rule R for two configurations $c_1 = \sigma_1(g_{11}, g_{21}, \ldots, g_{n1}), c_2 = \sigma_2(g_{12}, g_{22}, \ldots, g_{n2})$ is given $c_1 R c_2$ iff there is a permutation $(1, 2, 3, \ldots, n) \leftrightarrow (i_1, i_2, i_3, \ldots, i_n)$ that maps generators and connections from c_1 to c_2. Hence $content(c_1) = content(c_2)$ and the topology of connectors is preserved. In other words, the image is the invariant set under the group of graph automorphisms.

---

[2] See GPT p. 43 and p. 106.
[3] See GPT, Section 2.1 concerning identification rules.

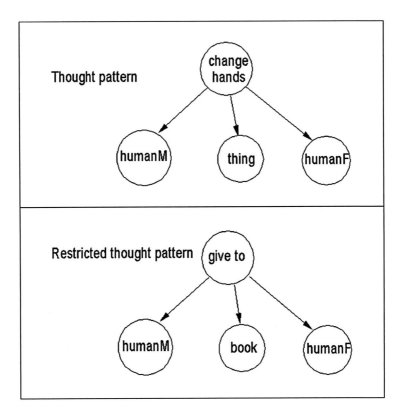

Figure 2.5:

It is known that the general graph isomorphism problem is computation-ally demanding although perhaps not NP-complete. In the present context, however, we are dealing with a more restricted problem where computing may not be overwhelming, see Jean-Loup Faulon (1998).

A partial characterization of thoughts is through the *M-ness ratio*. For a given $thought = \sigma(g_1, g_2, \ldots, g_n)$ and a collection $\mathbf{M} \subset \mathcal{M}$ we have the

DEFINITION: *The M-ness ratio of a thought is*

$$R_\mathbf{M}(thought) = \frac{|\{i : g_i \in \mathbf{M}\}|}{n}. \tag{2.11}$$

For example, with $\mathbf{M} = AGGRESSION$, shown in Figure 2.6.

The M-ness ratio can be used as an indicator to find the theme (to be introduced later) dominating the thought process at a certain time.

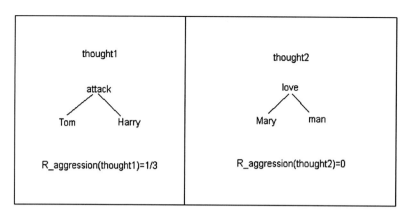

Figure 2.6:

2.5   Some Thoughts in Goethe

Let us now illustrate the construction by some thoughts appearing
in a famous novel by Goethe, "Elective Affinities", "Die Wahlver-
wandtschaften". This choice is especially appropriate since, when Goethe
wrote his work, he was strongly influenced by then current theories of chemistry based on affinities between substances, similar to the bonds between
ideas that we have postulated for human thought processes. We shall only
look at some simple thoughts and hope that some researcher will pursue
this attempt more fully.

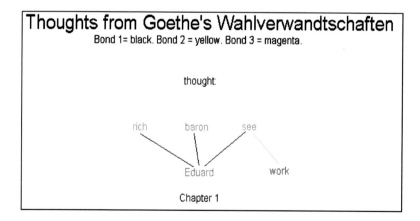

Figure 2.7:

A simple example is Figure 2.7 with the interpretation: "the rich baron Eduard sees work" and another simple one shown in Figure 2.8 with the interpretation: "the gardener answers Eduard that the place is new".

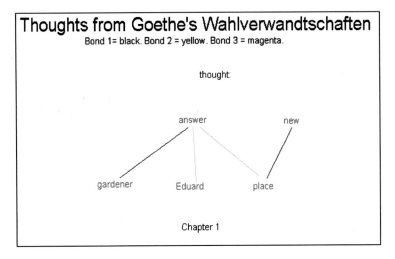

Figure 2.8:

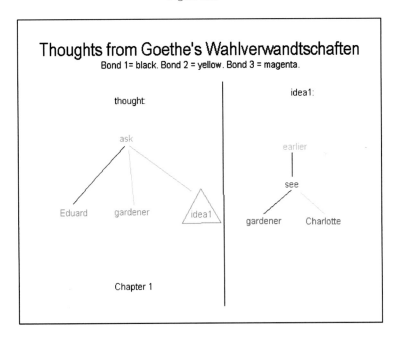

Figure 2.9:

The next one involves encapsulation of an idea, see Figure 2.9.

Interpretation: "Eduard asks the gardener something", something = "gardener has seen (someone) earlier". Recurrent thought with nested encapsulation is seen in Figure 2.10.

Interpretation: "Eduard says that Charlotte requires that she waits for him". The next three figures show slightly more complicated thoughts.

First, Figure 2.11 with the interpretation: "Eduard follows the gardener who walks away fast".

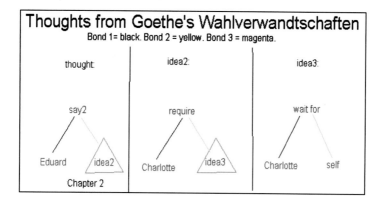

Figure 2.10:

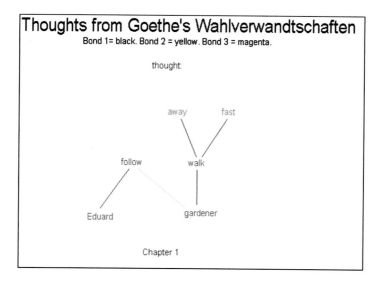

Figure 2.11:

Then Figure 2.12:

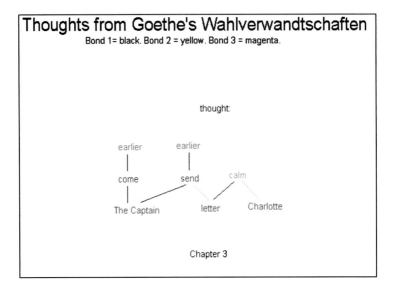

Figure 2.12:

Interpretation: "The Captain came earlier and sent earlier a letter to calm Charlotte". Finally Figure 2.13:

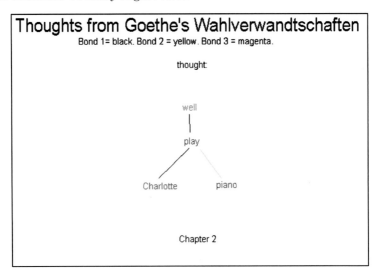

Figure 2.13:

Interpretation: "Charlotte plays the piano well".

Some of these examples show connected graphs, or, to use our terminology, they represent conscious thoughts. This is the result of thought chatter, eventually resulting in a dominating thought. Chatter may look like Figure 2.14.

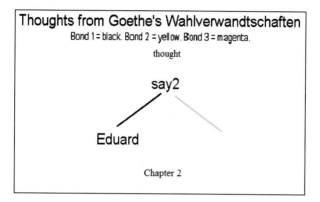

Figure 2.14:

Note that bond No. 2 from "say" (in yellow) is not connected. If the thought chatter had been allowed to continue it may have led to a complete thought. Figure 2.15 illustrates how Goethe makes a generalization, using A, B, C, D as modalities.

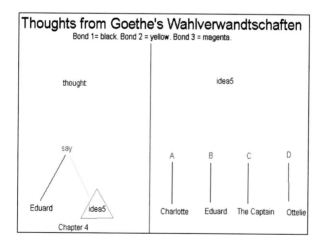

Figure 2.15:

35

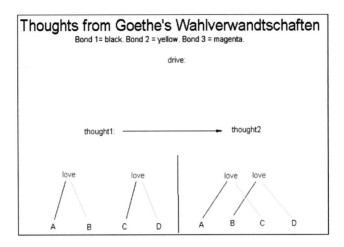

Figure 2.16:

Interpretation: "Eduard says idea5", with idea5 = "let the modality A contain Charlotte, the modality B contain Eduard, ...". Next, Figure 2.16.

Interpretation: the driver "thought1 to thought2" with thought1 = "A loves B and C loves D"; thought2 = "A loves C and B loves D". It actually represents a thought transformation with a composite move, a double romantic relation changes into another. Or, see Figure 2.17.

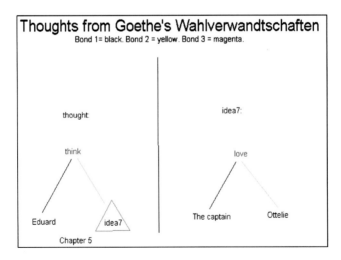

Figure 2.17:

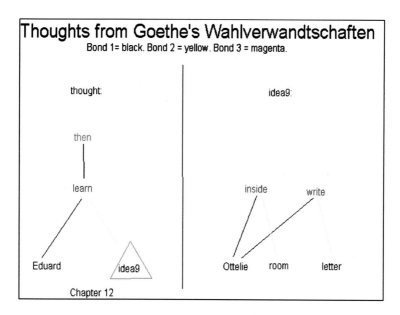

Figure 2.18:

Interpretation: "Eduard thinks that idea7" with idea7 = "The Captain loves Ottelie". Another abstraction with a new idea9 is seen in Figure 2.18.

Interpretation: "Eduard then learns that Ottelie is inside the room writing a letter".

Enough of Goethe. We hope that other researchers will expose GOLEM to other literary or artistic environments, so that it can develop more rounded personalities. That is for later.

Part II
Personality of a Mind

Chapter 3

A Mind Equation

In this section we shall limit ourselves to a simple mind, incapable of abstractions and generalizations and not subject to inputs from the external world. In later sections these restrictions will be removed.

We have seen how the set of all regular thoughts, complete and incomplete, constitutes the MIND. It represents all the thoughts that are possible currently, whether likely to occur or not. For a particular individual its MIND may change over time, modifying the idea space G, but momentarily we shall not let the MIND be capable of producing any new elementary ideas. That does not mean that all thoughts in the MIND are equally likely to occur. On the contrary, some will occur more often than others: due to external stimuli and remembered events, some are more likely than others. To formalize this we introduce an *idea function* Q taking positive values over the generator space, $Q(g) > 0$; $g \in G$. A large value of $Q(g)$ means that the elementary idea g is likely and vice versa. The Q-values need not be normalized to probabilities, for example $Q \equiv 1$ is allowed and means no preference for any generator.

So a person overly concerned about his wealth will have large values for $Q(money)$, $Q(stocks)$, $Q(rich)$, $Q(acquire)$, ..., while someone more concerned about physical appearance will emphasize $Q(looks)$, $Q(Vogue)$, $Q(mascara)$, ... As the situation changes from one genre to another the Q-function will change; more about this later.

But the specification of the Q-function does not describe how one simple idea is likely to associate with another. To do this we introduce a positive *acceptor or association function* $A(g_1, g_2)$: a large value of $A(g_1, g_2)$ means that the ideas g_1 and g_2 are likely to be associated (directly) with each other in the thinking of MIND and vice versa; see GPT, Chapter 7.

We shall now build a probability measure P over MIND and base the construction on the following:

RATIONALE: For a regular thought regular $c = thought \in MIND = \mathcal{C}(\mathcal{R})$

$$thought = \sigma(g_1, g_2, \ldots, g_n), \tag{3.1}$$

we shall assume its probability $p = P(thought)$

1) to be proportional to $Q(g_i)$ for each node i in σ;

2) to be proportional to $A[b_j(g_i), b_{j'}(g_{i'})]$ for each connected pair of bonds $b_j(g_i), b_{j'}(g_{i'})$ in σ;

3) to be proportional to κ_n where κ_n is a complexity parameter for size n of *thought*.

To illustrate this consider the following picture with five elementary ideas "apple", "sweet", "sour", "good" and "bad"" with Q and A values as indicated.

We can then calculate the probabilities for thought

$P(apple, sweet, good) \propto Q1 \times Q2 \times Q4 \times A12 \times A24 = 18;$
$P(apple, sweet, bad) \propto Q1 \times Q2 \times Q5 \times A12 \times A25 = .12;$

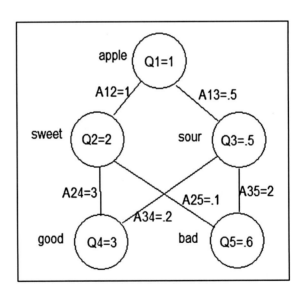

Figure 3.1:

$$P(apple, sour, good) \propto Q1 \times Q3 \times Q4 \times A13 \times QA34 = .15;$$
$$P(apple, sour, bad) \propto Q1 \times Q3 \times Q4 \times A13 \times A35 = .3;$$

so that the thought $\sigma(apple, sweet, good)$ is much more likely than the others.

Now we will make this more general. A reader who wants to avoid the mathematical specification can skip the formulas in the next few pages; the essentials are actually quite simple as in the above example.

We shall postulate the *mind equation*:

$$p(thought) = \frac{\kappa_n}{n! Z(T)} \prod_{i=1}^{n} Q(g_i) \prod_{(k,k') \in \sigma} A^{1/T}[b_j(g_i), b_{j'}(g_{i'})]. \qquad (3.2)$$

In this formula we multiply Q and A values, just as in the example above, and with bonds represented by the coordinates $k = (i,j), k' = (i',j')$, edges (k,k') in the connector graph, a temperature T, and a partition function $Z(T)$. Recall that the i's are generator coordinates and the j's bond coordinates. The positive parameter T, the *temperature*, expresses the mobility of the mind: high temperature means a lively, perhaps unruly mind, and low temperature characterizes a rigid mind. The factor κ_n makes the probability depend upon the size n of the thought, so that a mind capable of complex thinking has a good deal of probability mass for large values of n. The mind equation is a special case of the second structure formula (see GPT p. 366).

REMARK. In MIND we shall often replace $A[b_j(g_i), b_{j'}(g_{i'})]$ in formula (3.2) by $A(g_i, g_{i'})$. In other words, we shall govern the creation of the random connector σ by what are the elementary ideas to be connected, not by what are their out-bonds. Then the acceptor function A is defined on $G \times G$ instead of on $\mathcal{M} \times \mathcal{M}$. It should be clear from context which is the case.

In order that the mind equation make mathematical sense it is necessary that the κ_n decrease fast enough, preventing infinite thoughts to occur. Precisely how this is done is proven in Appendix A where the condition takes the form $\kappa_n = O(\rho^n)$ where ρ is less than a certain constant. Later, and in the software for GOLEM, we shall assume that $\kappa_n = \rho^n$.

The bonds take values depending upon what mind modality *mod* a generator belongs to. A generator $g \in mod \subset \mathcal{M}$ with arity ω will have out-bonds $b_j(g); j = 1, 2, \ldots, \omega(g)$ and all in-bonds equal to *mod*. Note that the *connector σ in* (3.2) *is variable and random* which motivates the appearance of κ_n which controls how likely are thoughts of different complexities.

The factor $n!$ means that permutations of the connector graph with its generators would have no real effect. It will sometimes be convenient to work with energies q, a instead of Q- and A-functions

$$Q(g) = exp[-q(g)]; \quad A(b_1, b_2) = exp[-a(b_1, b_2)]. \tag{3.3}$$

Then the energy of a thought can be written as

$$E(thought) = log(n!) - log(\kappa_n) - \sum_{i=1}^{n} q(g_i)$$
$$- 1/T \sum_{(k,k')\in\sigma} a[b_j(g_i), b_{j'}(g_{i'})]. \tag{3.4}$$

Here we have left out the term corresponding to the partition function Z; energies are determined up to an additive constant so that we have just normalized the expression for convenience. It has to be reintroduced when we use the relation $E = -log(p)$.

If two bonds $k_1 = (i_1, j_1)$, $k_2 = (i_2, j_2)$ have an interaction energy that is positive, $a(k_1, k_2) > 0$, the bond couple is *repellent*, the bonds are unlikely to close. On the other hand if $a(k_1, k_2) < 0$, *attractive bonds*, the bond couple is more likely to close. Note that open bonds are not automatically likely to close, it depends upon whether the couple is repellent or attractive.

More precisely we have the following

PROPOSITION: *For a thought $T = (T_1, T_2)$ consisting of two independent thoughts (see later) we have the energy relation*

$$E(T) = E(T_1) + E(T_2) + \binom{n_1 + n_2}{n_1}. \tag{3.5}$$

PROOF: We have, using the geometric series form of κ_n,

$$E(T1) = log(n_1!) - n_1 log(\rho) - Q1 - A1, \tag{3.6}$$
$$E(T2) = log(n_2!) - n_2 log(\rho) - Q2 - A2, \tag{3.7}$$
$$E(T) = log(n!) - n log(\rho) - Q - A, \tag{3.8}$$

where the Q's and A's mean the respective sums in equation (3.4) and n_1, n_2, n are the sizes of the thoughts T_1, T_2, T. Then

$$E(T) = E(T_1) + E(T_2) + log(n!) - log(n_1!) - log(n_2!) \tag{3.9}$$

which reduces to the stated result in (3.5).

Hence, the energy for independent thoughts is additive except for a term $log[B(n_1, n_2)]$, the log of a binomial coefficient. Since binomial coefficients are always bigger than (or equal to) one, it follows that energy is super-additive. Combining thoughts demand more and more mental power as the sizes increase: the MIND is limited in the complexity of thoughts.

We should think of Q as specifying the constitution of the *mental soup* in which the MIND is immersed at the time. This soup will depend upon the external world that we shall later characterize in terms of themes (or genres). This is likely to change during over time. It will also depend upon internal characteristics that may be more persistent, the personality profile, to be treated later.

The Q's and A's determine the *character of an individual mind*: two minds, MIND1 and MIND2, can have the same mental potential, MIND1 = MIND2, but different characters, same competence but different performance to borrow Chomsky's terminology.

It should be pointed out that the probability measure defined by the structure formula can be an adequate description of the mental activities only when MIND is at rest, not subject to specified input from the external world and not conditioned by any fact requiring attention: we are dealing with uncontrolled thinking. Otherwise P must be modified; this will be discussed in depth later. To distinguish the above P from more intelligent ones we shall call it the probability measure of *free associations*.

This defines the configuration space $\mathcal{C}_{complete}(\mathcal{R})$ consisting of all complete thoughts and the configuration space $\mathcal{C}(\mathcal{R})$ that also includes incomplete thoughts.

3.1 Personality Profile

Each MIND has a *self* $\in G$. The behavior of "self" is regulated by personality parameters *greedy, scholastic, aggressive, selfish,* .... The values of the parameters are in the interval (0,1) so that for example "generous" controls the value of $A(self, g)$ with "g" = "give", "lend", ... Their values constitute a personality profile that remains fixed after having been set.

The concept of *personality* should be compared to that of "genre" (or "theme") which can vary quickly over time and controls the values of "Q". The genre is not associated with any "self"; it describes the current *mental environment*.

3.1.1 *The Pathological Mind*

So far we have been studying a normal mind with no pronounced abnormalities. In the opposite case, however, the mind parameters including Q and A but also channel properties (to be discussed later) will have to be modified considerably. For example, a bipolar personality will have Markovian transition intensities take more pronounced values than for the normal MIND. Or, in the case of depression, the Q values may be large for the modalities MOOD, AFFECT, FEELING2 and EMOTION2, while in a schizoid situation the modalities ILLUSION, MEMORY1, MEMORY2 and AFFECT can be emphasized. We have been very sketchy here and have not really tried to examine these possibilities; it must be left to others, more qualified for this purpose, to try out alternative specifications of the MIND.

Before we leave this subject, however, we shall take a brief look at how such mind specifications could be the basis for testing procedures. Say that we have specified for normal and abnormal Q_{normal}, A_{normal} and $Q_{abnormal}, A_{abnormal}$ respectively for some particular mental disorder. The resulting probability measures will then be given by equations (3.10) and (3.11).

$p_{normal}(thought)$

$$= \frac{\kappa_n}{n! Z_{normal}(T)} \prod_{i=1}^{n} Q_{normal}(g_i) \prod_{(k,k') \in \sigma} A_{normal}^{1/T}[b_j(g_i), b_{j'}(g_{i'})], \quad (3.10)$$

$p_{abnormal}(thought)$

$$= \frac{\kappa_n}{n! Z_{abnormal}(T)} \prod_{i=1}^{n} Q_{abnormal}(g_i) \prod_{(k,k') \in \sigma} A_{abnormal}^{1/T}[b_j(g_i), b_{j'}(g_{i'})].$$

$$(3.11)$$

Given a set $THOUGHT = (thought_1, thought_2, \ldots, thought_t, \ldots, thought_m)$ of observed thoughts in a MIND, with $thought_\sigma^t(g_1^t, g_2^t, \ldots, g_{n^t}^t)$; $t = 1, 2, \ldots, m$ and we want to test the hypothesis H_{normal} against another hypothesis $H_{abnormal}$. Proceeding according to the Neyman-Pearson scheme we would calculate the likelihood ratio

$$L = \frac{Z_{abnormal}}{Z_{normal}} \prod_{t=1}^{m} \prod_{i=1}^{n^t} \frac{Q_{abnormal}(g_i^t)}{Q_{normal}(g_i^t)} \prod_{(k,k') \in \sigma^t} \frac{A_{abnormal}^{1/T}[b_j(g_i^t), b_{j'}(g_{i'}^t)]}{A_{normal}^{1/T}[b_j(g_i^t), b_{j'}(g_{i'}^t)]}.$$

$$(3.12)$$

The trouble with this expression is the occurrence of the two partition functions Z_{normal} and $Z_{abnormal}$ that are notoriously difficult, not to say impossible to compute. We can avoid this difficulty as follows, which also happens to make practical sense.

Observe thoughts in the set THOUGHT, where some may be incomplete, at least for the abnormal mind, and note how MIND completes them and probably adds new elementary ideas. The results will be a set $THOUGHT^{new} = (thought_1^{new}, thought_2^{new}, \ldots, thought_t^{new}, \ldots, thought_m^{new})$. Form the conditional probability under both hypotheses

$$P_{normal}(THOUGHT^{new} | THOUGHT)$$

$$= \frac{P_{normal}((THOUGHT^{new} | THOUGHT) \text{ and } THOUGHT)}{P_{normal}(THOUGHT)},$$

$$P_{abnormal}(THOUGHT^{new} | THOUGHT)$$

$$= \frac{P_{abnormal}((THOUGHT^{new} | THOUGHT) \text{ and } THOUGHT)}{P_{abnormal}(THOUGHT)}.$$

To evaluate these probabilities using equation (3.2) we note that the partition function appears in both numerator and denominator and hence cancel so we do not need them. Also note that $((THOUGHT^{new} | THOUGHT) \text{ and } THOUGHT) = THOUGHT^{new}$ since $THOUGHT_{new}$ includes $THOUGHT$.

REMARK. In a similar way we could handle the situation when MIND is also allowed to delete elementary ideas in THOUGHT but we shall not go into this here.

To continue evaluating the conditional probabilities we shall cancel factors occurring in both numerator and denominator. This gives us the simpler expressions

$$P_{normal}(THOUGHT^{new} | THOUGHT)$$

$$= \prod_{t=1}^{m} \prod_{added} Q_{normal}(g_i) \prod_{(k,k') \in \sigma_{(added)}} A_{normal}^{1/T}[b_j(g_i), b_{j'}(g_{i'})],$$

$$P_{abnormal}(THOUGHT^{new} | THOUGHT)$$

$$= \prod_{t=1}^{m} \prod_{added} Q_{abnormal}(g_i) \prod_{(k,k') \in \sigma_{(added)}} A_{abnormal}^{1/T}[b_j(g_i), b_{j'}(g_{i'})].$$

In these expressions the notation "added" means that the products over i and (k, k') should only extend over the values belonging to the new elementary ideas in $THOUGHT_{new}$. Then we get a test for abnormality:

PROPOSITION: *The critical region W[1] for testing abnormality by a most powerful test is given by the inequality*

$$W = \left\{ \prod_{t=1}^{m} \prod_{added} \frac{Q_{abnormal}(g_i)}{Q_{normal}(g_i)} \prod_{(k,k') \in \sigma_{(added)}} \frac{A_{abnormal}^{1/T}[b_j(g_i), b_{j'}(g_{i'})]}{A_{normal}^{1/T}[b_j(g_i), b_{j'}(g_{i'})]} \right.$$
$$\left. > const \right\}. \tag{3.13}$$

Say now that we have observed the MIND's reaction to a set $TEST = (test_1, test_2, \ldots, test_s, \ldots, test_r)$ of test thoughts given externally in the same way as will be done in GOLEM in the mode THINKING DRIVEN BY EXTERNAL INPUTS. Notice that these thoughts are not generated by MIND itself but by someone else, the analyst. The MIND will respond with some thoughts that we shall denote as above with the set $THOUGHT_{new}$.

How can we design a test of mental disorder in such a situation? Then it does not seem possible to eliminate the influence of the partition function, at least not with the above device. Perhaps some reader will be able to come up with a reasonable test. Perhaps one could derive a probabilistic limit theorem for stochastic variables of the form

$$\sum_i q(g_i) + \sum_{(k,k')} a(b_j(g_i), b_{j'}(g_{i'})) \tag{3.14}$$

and use it to get an approximate critical region. But we are digressing — let us return to the normal mind.

3.1.2 *An Intelligent Mind?*

A mind that deserves to be called intelligent must be able to handle complex ideas, for example the way three simple ideas combine to give rise to a new one. This is related to the celebrated Hammersley-Clifford theorem, see Hammersley-Clifford (1968), which says that on a fixed, finite graph σ with assigned neighborhood relations a probability density p is Markovian iff it

---

[1] See Cramer, Section 35.3.

takes the form

$$p = exp[-E(c)]; \quad E(c) = \sum_{cliques \subset \sigma} E_{cliques}(g_1, g_2, \ldots, g_r). \quad (3.15)$$

The sum is over the cliques of σ. A clique is a subset of the graph all whose sites are neighbors in the topology of σ. Note, however, that this theorem does not apply without modification to our situation, since the connector graphs we are dealing with are not fixed but random. Anyway, it gives us a hint on how to organize a more powerful mind.

Instead of using only functions of a single generator, like $Q(g)$, or of two, like $A(g_1, g_2)$, we are led to use energies that depend upon more than two generators. In other words, the mind is controlled by a randomness that involves ideas of higher complexity than size 2. For the specification of P in the previous section we could let the acceptor function depend upon the triple $\{man, love, woman\}$, not just on the pairs $\{man, love\}$ and $\{woman, love\}$.

Having said this, it should be pointed out that this increase in mental complexity could also be achieved by increasing the generator space as described in GPT, Section 7.3, that is by forming macrogenerators by combining the original generators. Which of these two procedures we should choose is a matter of convenience and practicality, not of principle: are we most concerned with keeping the cardinality of the generator space manageable or with dealing with small dimensions of energy functions? Whichever alternative we choose, we extend the intellectual power of the synthetic mind. In the code GOLEM we shall choose the latter alternative.

3.1.3 *Randomness and Thinking*

We emphasize that thought processes must include random elements, we do not consider them deterministic. This difficulty cannot be avoided, randomness is forced upon us. A purely deterministic, completely rigid, theory of mind is doomed to fail.

3.2 Mental Dynamics

The above was dealing with the mind at rest, a static system. Now let us consider the development in time.

3.2.1 *Topologies of Thinking*

We need a concept "near" for thoughts: one thought may be close to another thought but not to a third one, and therefore we introduce neighborhoods $N(thought)$, in mind space by

$$N(thought)$$
$$= \{\forall \; thought' \ni thought' \; and \; thought \; differ \; only \; in \; one$$
$$generator \; or \; one \; connection\},$$

similar to the discussion in GPT, Section 5.2. This imposes a topology on both $\mathcal{C}_{complete}(\mathcal{R})$ and $\mathcal{C}(\mathcal{R})$, formalizing the concept "thoughts close to each other".

With such topologies it makes sense to talk about *continuity of thought* (although with a discrete interpretation) and *jumps in thinking*, which will be done when discussing the algorithms giving rise to trajectories in MIND space, one thought leading to another. In particular, composite moves. The trajectory will prefer to climb hills in the probability landscape as in Figure 3.2 or over a longer durations, see Figure 3.3.

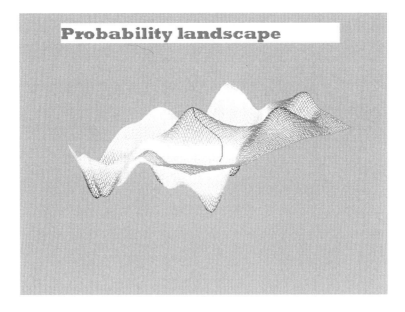

Figure 3.2:

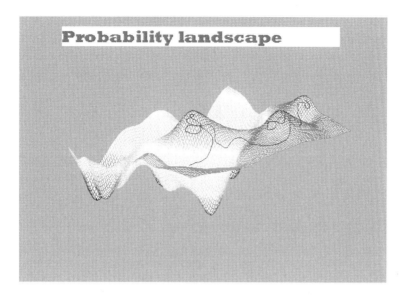

Figure 3.3:

3.3 Personality Profiles in Literature

The diagrams in the previous section illustrate a thought process — but whose thoughts? Certainly not Ottelie's nor the Captain's. It seems more convincing to attribute them to Goethe himself during the period while he was writing "Elected Affinities". If we do this then the diagrams express Goethe's personality, at least to some extent.

This leads us to the question whether personality analysis in literature can be organized in the same way in general, and what would we gain in understanding by such an approach. In literature studies a related idea is *content analysis*, see for example Holsti (1969) and above all the pioneering work of Propp (1927) in which Russian folk tales were analyzed to produce a morphology in terms of simple entities like, actors, actions, environment and so on, remarkably like the fundamentals in Pattern Theory. But how should such a task be organized?

It is similar to automatic language translation but the latter concerns transformation from one language to another, transforming one structure to another one like it. Our present task deals with transformations from language to a MIND-structure with rather different properties. First we would need a large collection of ideas, organized into a generator space G

with arities, modalities and bond values. In other words, a transformation from short language strings to idea content in G. This is a tall order! Perhaps one could start from something like WordNet, see References. But this is not enough. Indeed, we also have to transfer the meaning of various linguistic devices, for example

"going to" to future

"not very" to negative, quantity

"do not" to negative, action

to mention but a few. In addition we need a filter that selects the important, information carrying, words. Such words as "the", "but" and "a" do not support much information and could be neglected. This could be compared to the remarkable study in authorship determination Mosteller-Wallace (1964) of the Federalist Papers. Those authors argued that words that carry specific information like "presidential", "federal", "voting" should not be used to separate hypothesis about authorship since they are too "special". Instead "politically innocent" words like "whose" and "there" should be preferred. But our task is different, it is not to characterize linguistic style but modes of thinking in whatever format they are expressed. This task presents a formidable difficulty but it may not be insurmountable given a massive research effort.

Then we would need an algorithm that computes a "likely" (optimal) connector joining the transformed elements of G to regular thought expressions. Perhaps "likely" should be interpreted in terms of a preliminary personality profile Q^*_{pre}, A^*_{pre}. To achieve an optimal construction we would probably use a greedy algorithm.

Third, we would estimate the personality profile Q, A successively as we process the text. Such estimation problems have been dealt with in GPT, Chapter 20.

In any serious attempt to tackle this problem new perspectives would appear, necessitating a reappraisal of the task.

3.4 Trajectories in Mind Space

But how can we compute the probabilities of possible thoughts in $MIND = \mathcal{C}(\mathcal{R})$? In particular, how can we avoid the computation of the infamous partition function? This will be accomplished by a variation of *stochastic relaxation*, see GPT p. 379. The main trick in this celebrated technique is to exploit the Markovian nature of the measure P over mind space (not to be confused with the fact that stochastic relaxation produces a chain that is Markovian over time).

Actually, we need not compute the probabilities of possible thoughts; instead we shall *synthesize* the random mental states by an iterative procedure where each step consists of a *simple move,* or later a composite move, through mind space. This technique is well known to practitioners of MCMC, Monte Carlo Markov Chain.[2] A difference to the usual way one applies MCMC, however, lies in the fact that for mind representations *the connector is also random,* not just the generators at the sites of a fixed graph. To develop a mental dynamics of the mind we shall think of a trajectory through mindscape, through MIND, as made up of steps, usually small, but occasionally bigger. Among the simple moves that we have in mind we mention only a few here:

(1) Place a generator at a new site; no new connections will be established in this move.

(2) Delete a generator in the *thought* and the connections to it. This step automatically respects regularity since the regular structure $MIND = \mathcal{C}(\mathcal{R})$ is monotonic.[3]

(3) Delete a connection in σ; also respects regularity (but not complete regularity).

(4) Create a connection between two generators in *thought* if regularity allows this.

(5) Select a generator $g \in thought$ and replace it by another one g' including the possibility of keeping it unchanged, observing the regularity constraint $mod(g) = mod(g')$.

All of these moves represent *low level mental activity,* for example the transformations $dog \to dog, big$ and $man \to man, walk$. For each of them we define a random selection rule for choosing among the possible alternatives allowed by the regularity constraints.

REMARK. It should be observed that such simple moves actually map thoughts to *sets* of thoughts when the randomness of the transformation $\mathcal{T}$ is taken into account:

$$\mathcal{T} : MIND \to 2^{MIND}. \tag{3.16}$$

---

[2] See GPT, Chapter 7.
[3] See GPT, p. 6.

But how do we randomize these choices so that we get the desired probability measure?

To do this it is important to select the set $\mathcal{T}$ of moves, $T \in \mathcal{T}$, sufficiently big. More precisely, in order that they generate an *ergodic* Markov chain, which is required for the following argument, it is necessary for any pair of regular configurations $c_1, c_N \in \mathcal{C}(\mathcal{R})$ that there exist a chain $c_2, c_3, \ldots, c_{N-1}$ and $T_1, T_2, \ldots, T_{N-1}$ such that

$$c_2 = T_1 c_1, c_3 = T_2 c_2, \ldots, c_N = T_{N-1} c_{N-1}; c_i \in \mathcal{C}(\mathcal{R}) \text{ and } T_i \in \mathcal{T} \ \forall i. \tag{3.17}$$

In other words: any thought in *MIND* can be continued to any other thought by a sequence of thoughts, one close to the next one. The chain may be long but finite. This makes the Markov chain (over time) irreducible and since we shall make it have P as an equilibrium measure, it follows[4] that the chain is ergodic. The importance of ergodicity was emphasized in the research program described in the CD-ROM "Windows on the World". It guarantees that the mind is not too rigid so that it is possible to pass from any mental state to any other. We shall assume that this is so in the following.

REMARK. On the other hand it may be of some interest to also study situations when the mind is not ergodic, so that it is constrained to a proper subset of MIND. Such a mind just cannot realize transitions between certain thoughts and emotions that would otherwise be consistent with the mental setup, it is abnormally restricted. Therefore the importance of ergodicity is clear. The fact that the Markov chain is irreducible guarantees that *the mind is not too rigid*, so that it is possible to pass from any mental state to another. Otherwise it can be caught constrained to a part of *MIND*, not being possible to exit to other (possible) mind states.

The above applies to fairly short time intervals, say minutes and hours, during which time the MIND has not had time to modify its parameters, G, Q, A substantially. However, for longer durations *the MIND is an open system*, successively modified due to new experiences and input from the surroundings. Also creating new ideas as we shall discuss later. Then ergodicity does not apply.

On the other hand, when we deal with associations that are not free but dominated by attention to some theme, we shall make the mind almost

---

[4]See Feller (1957), Section XV.6.

non-ergodic: the probability of reaching outside a given theme will be close but not equal to zero.

As the generators and/or connections are being changed successively we get a trajectory in mind space

$$thought_1 \to thought_2 \to thought_3 \ldots \qquad (3.18)$$

which represents *a train of thoughts*, some conscious, others not, a trajectory through mental domain *MIND*. The intermediate thoughts play the role of the links in Poincare's chain of thought.

The reader may want to avoid the technicalities in the next section by skipping it.

3.5   Dynamics with Simple Moves

Let us still deal with a situation when no external stimuli impact on the mind and where the time duration is so short that we can neglect changes in the mind energies q and a.

Let us explain the way we make use of the Markovian nature of P. Say that we are dealing with a transformation $T : MIND \to MIND$ that only affects a single generator g_i at site $i \in \sigma$, see Figure 3.4.

The site i has the neighbors $2, 4, 10, 11$ so that we can write the conditional probability

$$P(g_i|g_1, g_2, g_3, g_4, g_5, g_6, g_7, g_8, g_9, g_{10}, g_{11})$$

$$= \frac{P(g_i, g_1, g_2, g_3, g_4, g_5, g_6, g_7, g_8, g_9, g_{10}, g_{11})}{P(g_1, g_2, g_3, g_4, g_5, g_6, g_7, g_8, g_9, g_{10}, g_{11})}.$$

But we can use (3.2) to reduce this expression by canceling common factors in the numerator and denominator, leading to

$$P(g_i|g_1, g_2, g_3, g_4, g_5, g_6, g_7, g_8, g_9, g_{10}, g_{11})$$
$$= \frac{P(g_i, g_2, g_4, g_{10}, g_{11})}{P(g_2, g_4, g_{10}, g_{11})}.$$

This simplification is not very useful for thoughts consisting of just a few generators, but if the number is large, it amounts to a considerable gain in computing effort.

TOPOLOGICAL ENVIRONMENT IN CONNECTOR GRAPH

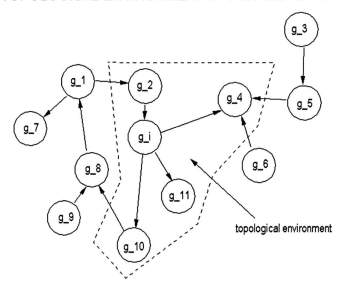

Figure 3.4:

In this way we can express the conditional probabilities we need for stochastic relaxation in the form

$$P(A|B) = \frac{N}{D} \tag{3.19}$$

where N and D are joint probabilities of sets in $\mathcal{C}(\mathcal{R})$ of moderate dimension. This reasoning was for simple moves involving only changes of generators while leaving the connector σ unchanged. If the connections in the connector also can change, they have to be included among the variables that make up the sample space of the relaxation procedure. Then the topology induced by the neighborhood relations has to be adjusted in the obvious way, but the general procedure remains the same as just described.

We choose a set of configuration transformations $\mathcal{T} = \{T^1, T^2, \ldots, T^\nu\}$, for example $\mathcal{T} = \{(2), (5)\}$, see the last section. It is large enough to span MIND, and we shall now construct updating algorithms for each T^l.[5] Apply the transformation $T = (2)$, with deletion at site m or no deletion at

---

[5]See e.g. GPT, Section 7.6.

all with given probabilities, to the configuration $thought_{old}$ resulting in $thought_{new} = thought_{old}$. We need the probability for the new mental state which, using (3.2), is proportional to N/D with the numerator

$$N = \pi_{n-1} \prod_{i=1, i \neq m}^{n} Q(g_i) \prod_{(k,k') \in \sigma^m} A^{1/T}[b_j(g_i), b_{j'}(g_{i'})] \qquad (3.20)$$

where σ^m is the graph obtained from σ of *thought* by deleting the site m as well as bonds emanating from it. Similarly, the denominator is

$$D = \pi_n \prod_{i=1}^{n} Q(g_i) \prod_{(k,k') \in \sigma} A^{1/T}[b_j(g_i), b_{j'}(g_{i'})]. \qquad (3.21)$$

This gives us

$$N/D = \frac{\pi_{n-1}}{\pi_n Q(g_m) \prod_{(k,k') \in \sigma^-} A^{1/T}[b_j(g_i), b_{j'}(g_{i'})]} \qquad (3.22)$$

where σ^- means the graph consisting of the site m together with the bonds emanating from it. This we do for $i = 1, 2, \ldots, n$ as well as for no deletion in which case (3.22) should be replaced by $N/D = 1$.

REMARK. If global regularity requires that deletion of a site also requires the deletion of other sites and their bonds, then (3.22) has to be modified accordingly.

Now $T = 5$. For an arbitrary generator $g \in G$ we need the equivalent of (3.2) placing g at a site with modality $mod(g)$ or not introducing any new generator at all, so that

$$N/D = \frac{\pi_{n+1} \pi_n Q(g) \prod_{(k,k') \in \sigma^+} A^{1/T}[b_j(g), b_{j'}(g)]}{\pi_n} \qquad (3.23)$$

where σ^+ is the graph consisting of the new generator g and bonds emanating from it. Note that in general there are several ways of connecting g to the old configuration and (3.23) must be evaluated for all these possibilities. For the case of no change, the right hand side of (3.23) should be replaced by 1.

The stochastic relaxation then proceeds by iteration as follows.

Step $T = 1$: Compute the expression in (3.22) for $m = 1, 2, \ldots, n$, normalize them to probabilities and simulate deletion at site m or no deletion. Get the new *thought*.

Step $T = 5$: Compute the expression in (3.23) for this T, normalize and simulate. Get the new *thought*.

Step $T = 2$: ....

And continue until sufficient relaxation is believed to have been obtained. As in all applications of stochastic relaxation it is difficult to give precise advice about when this has been achieved. Trial and error may have to suffice.

The above development of thoughts, the *thought chatter*, is thus essentially random. Of course not purely random but controlled by the regularity constraint as well as by the mind parameters Q, A. This is reminiscent of chemical reactions: many reactions (thought developments) are possible, but only a few actually take place. For example the thought $(green \rightarrow cat, grass)$ is regular but has low probability. A reaction would probably result in the thought $(cat, green \rightarrow grass)$ which has higher probability, lower energy and would stay conscious for a while. The first, unlikely one, will only be glimpsed consciously, if at all, and be hidden in the thought chatter.

3.6   Mental Dynamics with Composite Moves

With the above set up only changes at a single site or at a single connection are made at each instance of a train of thought; the mental associations are simple in the sense that only short steps are taken in the mental trajectory space. The change in mind state only depends upon the neighboring states of mind. But we shall also allow *composite moves* where the changes involve larger sub-thoughts. We do not have in mind a strict cause and effect relation; we want to avoid determinism, so that we will continue to allow the changes to be random. The reason why we allow composite moves is not that it will speed up convergence to the equilibrium measure, which is the standard motivation behind similar devices in most applications of stochastic relaxation. Such speed up may indeed occur, but that is not our motivation. Instead we believe that the train of thought obtained by composite moves mirrors more closely what goes on in real thought processes. Of course we have no empirical evidence for this, only introspective observations.

REMARK. The version of stochastic relaxation we have used here is only one of many, actually the most primitive. In the literature several others can be found that are guaranteed to have faster convergence properties, but as mentioned, we are not concerned with speed here. Or are we? If our conjecture that thinking can proceed in large jumps is correct, it may be that this happens in order to speed up the thought process, omitting links in it that are known to the mind to be at least plausible. Worth thinking about!

Now let us mention some examples of composite moves. In Figure 3.5, the thought "dog black big" is transformed into "black big Rufsan" with probability .6, expressing the knowledge possessed by this mind that if a dog is black, it is most likely to be Rufsan, at least in some MIND.

Or, in Figure 3.6, meaning that a question often leads to a reply.

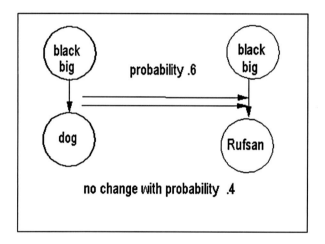

Figure 3.5:

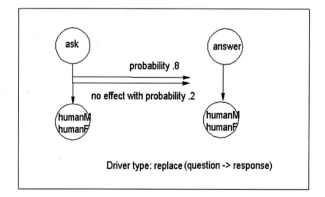

Figure 3.6:

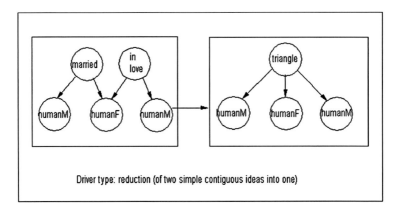

Figure 3.7:

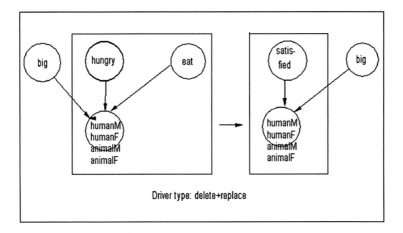

Figure 3.8:

Figure 3.7 describes how a thought with the five generators "humanM, humanF, humanM, married, in love" is transformed into the eternal triangle. In Figure 3.8 we see how hungry humans or animals will become satisfied after eating.

Some familiar drivers are shown in Figures 3.9–3.11.

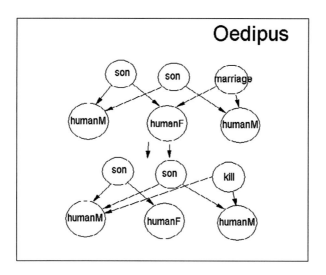

Figure 3.9:

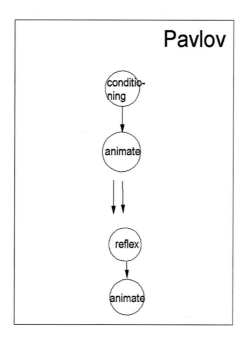

Figure 3.10:

Figure 3.9 shows the Oedipus complex, Pavlov's dog is in Figure 3.10, and Adler's self asserting individual is shown in Figure 3.11.

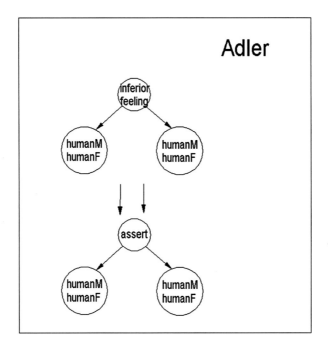

Figure 3.11:

The general form of a composite move is a transformation whose domain and range are sets of regular thoughts

$$Move : THOUGHT1 \rightarrow THOUGHT2;$$
$$THOUGHT1, THOUGHT2 \subset MIND, \qquad (3.24)$$

together with a probability measure $P_{move}, move \in MOVE$ over the set $THOUGHT1$. The measure P_{move} may be specified in the same way as for the simple moves, although their calculation will be more involved but it can also be modified to account for preferences of thinking. In this way the composite moves contribute to convergence to the equilibrium measure P just as the simple moves do, but the *trajectories will be different*, the steps $thought(t) \rightarrow thought(t+1)$ will be different, hopefully more realistic in characterizing the functioning of a particular mind. This applies to free associations. However, for less passive thinking the probabilities applied to composite moves may be different, influenced by attention to genres as will be discussed in the next section.

Note that we have implicitly allowed composite moves to apply to patterns of thoughts, not just to single thoughts.

We believe that a realistic mind representation will require many types of composite moves for the mind dynamics in contrast to static mind representation.

3.6.1 Mental Dynamics with Themes of Attention: Genres

Until now we have operated with a fixed equilibrium measure, P, but what happens when the mind genre changes? For example, when the domain of discourse concerns wealth and distribution of wealth. Or when the emphasis is on the emotional relation to another individual. To deal with such situations we shall let the Q-vector change, say by increasing the values of $Q(money)$, $Q(acquire)$, $Q(buy)$, $Q(sell)$, ... or $Q(love)$, $Q(jealousy)$, $Q(sweetheart)$, ..., so that the mind visits these ideas and their combinations more often than for free associations. Then the discourse is weighted toward a specific *genre* with a lower degree of ergodicity since it will take time to exit from these favored thoughts.

In this way we allow $Q = Q(t)$ to change in steps when one genre is replaced by another. We illustrate it in Figure 3.12; the circles represent constant Q and arrows indicate steps between mental genres. Different genres are connected via channels through which the mind passes during the thinking trajectory.

More formally, we introduce genres $\Gamma_r \subset G$ that are not necessarily disjoint, in terms of a-energies, and the *mind forces* F_r as the gradient vectors of dimension $|\Gamma_r|$ of the energies

$$F_r = (\ldots f_\mu \ldots); f_\mu = -\frac{\partial q}{\partial g_\mu}; g_\mu \in \Gamma_r. \tag{3.25}$$

This corresponds vaguely to the usage of "force" and "energy" (potential) in rational mechanics. This means that a force acts in the mind space to *drive* the mind into respective genres; it influences attention.

3.6.2 Mental Dynamics of Dreaming

To represent mind trajectories corresponding to dreaming and less conscious thought processes we shall make the binding between elementary

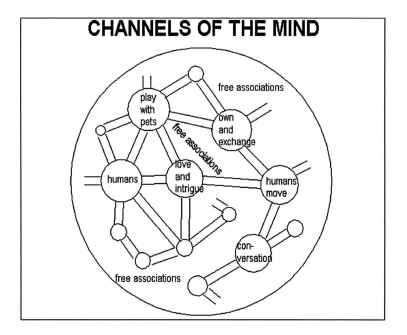

Figure 3.12:

thoughts less stringent, as dreams tend to allow strange and unusual transitions and associations. The technical way that we have chosen to do this is by increasing the temperature T appearing in the structure formula (3.12). A higher value for the temperature makes the value of the factor $A^{1/T}[b_j(g_i), b_{j'}(g_{i'})]$ closer to 1 so that the elementary thoughts, the generators, become less stochastically dependent (positively or negatively). In other words, the thinking becomes less systematic, and more chaotic.

3.7 A Calculus of Thinking

The MIND calculates. Not as a deterministic computer with strict algorithmic rules, but with a certain amount of controlled randomness. Among its algebraic operations, the *mental operations*, we mention especially two (more to follow):

$$\boxed{\textbf{mop1} = \text{SIMILAR: } \textit{thought} \mapsto s \textit{ thought}}$$

as illustrated in Figure 3.13:

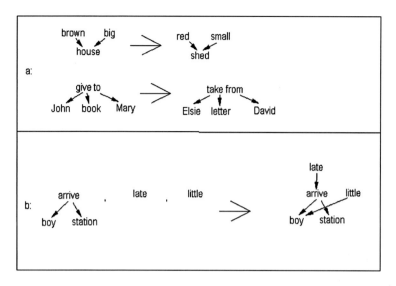

Figure 3.13:

and

$$\mathbf{mop2} = \text{COMPOSE: } thought1, thought2 \mapsto \sigma(thought1, thought2)$$

with some connector σ as illustrated in Figure 3.13b. We say that *thought1* contains the *thought2* if there exist a *thought3* such that *thought1* = *COMPOSE*(*thought2*, *thought3*).

Hence **mop1** changes a thought to one belonging to the same thought pattern, replacing elementary ideas with similar ones. The **mop2** combines two thoughts into a single one.

Note that this algebra is partial in that compositions of thoughts are only allowed if bond values agree in the coupling of the connector σ. The mental operations are formalizations of intuitive concepts of thinking processes. Approximate since the intuitive concepts are vague and not precisely defined. As all mathematical formalizations they increase precision but decrease generality.

With this architectonic approach, *pace* Kant, to the study of the mind, the most fundamental mind states, the elementary ideas, combine to make up the trajectories through the mind space *MIND*, governed by entities like Q, A, drives and so on. Certain regular sub-thoughts can be singled out because of their particular role. But how do we combine and operate on composite thoughts, how do we hook them together in Poincare's parlance? To do this we shall first consider some special instances.

3.7.1   *Specific Thoughts*

3.7.1.1   *Conscious Thoughts*

As the trajectory develops many associations are formed, most probably irrelevant. At a particular time t the total mind state $thought = thought(t)$ can be decomposed into connected components with respect to the topology induced by the total connector σ.

3.7.1.2   *Top-thoughts*

Another type of (sub)-thought is based on the notion of *top generator*.

   DEFINITION: *A top-thought in a total thought means a sub-thought (not necessarily a proper subset) that starts from a single generator and contains all its generators under it with respect to the partial order induced by σ. Its level is the level of its top generator. A maximal top-thought has a top generator that is not subordinate to any other generator in thought.*

   Let $tops(thought)$ denote the set of all generators in a *thought* that are not subordinate to any other generators. Then we get the decomposition

$$thought = top\_thought(g_1) \oplus top\_thought(g_2)\ldots; g_k \in tops \qquad (3.26)$$

where $top\_thought(g)$ stands for the sub-thought extending down from g. Note that in (3.26) the terms may overlap, two top-thoughts may have one or more generators in common as shown in Figure 3.14, where the two top-thoughts *idea*1 and *idea*3 in the lower part of the figure have the generator *John* in common but the top-thoughts above in Figure 3.13 do not: the latter can be said to be *regularly independent*: they are indeed independent as far as their meaning is concerned.

   Inversely, given two regular thoughts *thought*1 and *thought*2, we can form the composition

$$thought_{new} = thought_1 \overset{\sigma}{\oplus} thought_2; thought_1 = \sigma_1(g_{11}, \ldots, g_{1n_1});$$
$$thought_2 = \sigma_2(g_{21}, \ldots, g_{2n_2}) \qquad (3.27)$$

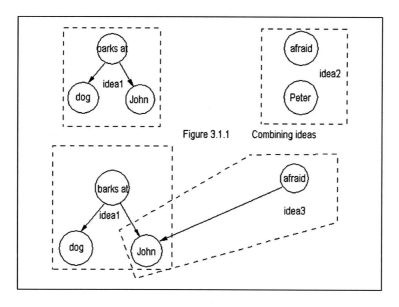

Figure 3.14:

where we have indicated by $\overset{\sigma}{\oplus}$ what generators, if any, $thought_1$ and $thought_2$ have in common; it can have the form

$$\sigma = \begin{cases} g_{1i_1} = g_{2k_1} \\ g_{2i_2} = g_{2k_2} \\ \ldots \ldots \end{cases} \tag{3.28}$$

If $thought$ is a top-thought, consider its external bonds

$$ext(thought) = ext_{up}(thought) \cup ext_{down}(thought) \tag{3.29}$$

consisting of up-bonds and down-bonds; note that all internal (i.e. closed) bonds are excluded.[6]

In Part III, when we start to build a mind, we shall be especially concerned with top-thoughts of level 2 although in general its level can be higher. This will lead to a mind that may be said to be intellectually challenged since its mental combination power is very restricted. We make this assumption only for simplicity; it ought to be removed.

---

[6] For a discussion of these concepts see GPT, Chapter 1.

3.7.2   *Generalization/Specialization Operation*

The process of generalization will here be understood in terms of the operator *MOD* that is first defined on $G \cup \mathcal{M}$ and takes a g into $mod(g)$ and a
modality m into itself. In the following it will be assumed that the modality
taxonomy is of Linnean form so that *MOD* is one-valued (it would however
be of interest to consider the case of non-Linnean taxonomy in which case
the generalization operator can be many-valued). It is then extended in the
natural way to $\mathcal{C}(\mathcal{R})$ by operating individually on each component. The
operator *MOD* is distributive over composition, so that *MOD*(*thought*) is
defined for any *thought* $\in$ *MIND*.

For example,

$$MOD(bark \downarrow Rufus) = (animal\_sound \downarrow animalM) \tag{3.30}$$

or

$$MOD(color \downarrow house) = (color \downarrow building). \tag{3.31}$$

The operator *MOD extends the meaning of a thought by suppressing inci-
dental information and hence deserves to be called generalization.* Hence the
mind calculus also has access to the operation

mop3 = GENERALIZATION: MOD TRANSFORM OF THOUGHT

It should be mentioned that the *MOD* operation can be iterated. For
example, we can get the successive generalizations *Rufsan* $\rightarrow$ *DOG* $\rightarrow$
ANIMAL_canine $\rightarrow$ *ANIMAL* $\rightarrow$ *ANIMATE*. What *generalization is useful* depends upon how often the thoughts contained in it will occur together.

But this deserves some comment. We have allowed modalities to join
in a limited way, combining parts of their contents that have common out-
bonds. Thus it makes sense to iterate the generalization operation **mop3**,
resulting in a *semi-group* $\mathbf{mop3}^{power}$; $power \in \mathbf{N}$. Actually, some reser-
vation is needed here to get a tree (or forest) structure. In the MATLAB
code for GOLEM only Linnean modality structure will be allowed. Anyway,
this makes it possible to form generalization of *thought* of the first order,
$power = 1$, of the second order, $power = 2$, and so on.

SPECIALIZATION

Figure 3.15:

The *specialization operation* does the opposite to generalization. In a $thought = \sigma(g_1, g_2, \ldots, g_n)$ one of the modalities m is replaced by $g \in m$.

3.7.3 *Encapsulation Operation*

Consider a *thought* $\in$ *MIND* with the top generator g_{top} on level l and $mod(g_{top}) = m$. If this *thought* occurs more than occasionally, the mind may create a new generator, a macro-generator, g_{macro} with the same interpretation as *thought* on level 1, up-bond *IDEA*, sometimes called *COMPLEX*. This *encapsulation procedure formalizes the mental process of abstraction*. Due to it the generator space has increased: the MIND can handle the idea as a unit with no internal structure.

For example,

$$thought = (married \downarrow humanM \ and \downarrow humanF) \qquad (3.32)$$

is abstracted to the macro-generator $g = marriage$ on level 1 with modality *IDEA*. Continuing the abstraction process we can introduce still another macro-generator *divorce* by abstracting the

$$thought = (dissolve \downarrow marriage) \qquad (3.33)$$

as *divorce* of modality *IDEA*. Hence the calculus also includes the operation

mop4 : ENCAPSULATION = ENCAPSULATED THOUGHT

Then we can consider a new thought as a unit,[7] a generator in the modality *IDEA*. This means a transformation

$$ENCAPSULATION : thought \rightarrow idea_k \in IDEA \subset G. \qquad (3.34)$$

We shall use many such generators in a modality called IDEA, often linked to generators like "say", "ask", "think". The transformation *ENCAPSULATION* plays an important role when representing mental states involving information transfer, for example

$$ENCAPSULATION : say \mapsto (black \downarrow Rufsan) \qquad (3.35)$$

with the right hand side as a generator in *IDEA* connected to *say*.

It should be mentioned that encapsulation can lead to configurations involving encapsulation again, *nested structures that allow the self thinking about itself* and so on. An iterated encapsulation *idea* will be said to have *power(idea) = p* if it contains p iterations. Once it is incorporated as a unit in G its power is reset to zero. This will have consequences for the updating of the memory parameters Q, A. More particularly, a new *idea* of size n, *content* $= (g_1, g_2, g_3, \ldots, g_n)$ and connector σ will be given a Q-value

$$Q(idea) = \frac{\kappa_n}{n!} \prod_{i=1}^{n} Q(g_i) \prod_{(k,k') \in \sigma} A^{1/T}[b_j(g_i), b_{j'}(g_{i'})] \qquad (3.36)$$

and A-values equal to one for those connections that are indicated by the modality transfer function and equal to a small positive number otherwise.

3.7.4 *Completion Operation*

If *thought* has some of its out-bonds left unconnected it will not be meaningful, it is incomplete. It can be made complete by adding elementary ideas so that all out-bonds become connected. This multi-valued operation is called COMPLETE, and in the software it is named DEEP THOUGHT since it

---

[7]This has been suggested in GPT, Section 7.3.

may require the MIND to search deeply for regular and hence meaningful extensions of *thought*. Or, symbolically,

| **mop5** : COMPLETE = DEEP THOUGHT |

3.7.5 *Genre Operation*

On the other hand, we can also change the probabilistic parameters that determine the behavior of MIND. Thus we have the GENRE operation

| **mop6** : genre: $Q \rightarrow Q_{genre}$; $genre \in GENRE$ |

3.7.6 *Inference Process*

Given the *thought* we can ask what the mind infers from it. This is done by another random mind operation

| **mop7** : INFER: *thought* $\rightarrow$ *thought*$_{infer}$ |

where *thought*$_{infer}$ is a random element sampled from MIND according to the *conditional probability relative to P that the element contains thought*. Note the way we have used conditioning of the probability measure to carry out inference. Actually, we use the term "inference" in a wider sense than what is standard. Usually "inference" means the process by which we try to interpret data in terms of a theoretical super-structure, perhaps using statistical methods. We shall, however, mean the mind process by which we extend a given thought, we *continue* it according to the probability measure P. Thus it is a random and multi-valued process.

From a given *thought* we can then infer a bigger one that naturally extends *thought* $\rightarrow$ *thought'*. For example, if $A(Rufsan, black)$ is big, we may get the inference *Rufsan* $\rightarrow$ *Rufsan, black*. This will happen if the MIND has seen the sub-thought *Rufsan, black* many times so that the memory updating has taken effect. On the other hand, we may not get the inference *black* $\rightarrow$ *black, Rufsan*, since it is unlikely that the MIND will select that inference from *black* from many others just as likely. This lack of symmetry seems natural for human thought.

$$\boxed{\textbf{mop8} \quad : \text{ MUTATE: } thought \rightarrow thought_{mutated}}$$

The mutation operation in it simplest form changes a generator g_i in $thought = \sigma(g_1, g_2, \ldots, g_n)$ into another g_i' belonging to the same modality, for example:

MUTATION

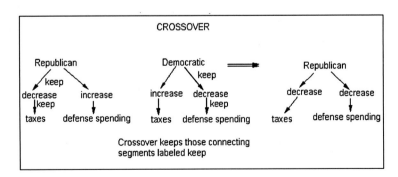

Figure 3.16:

However, a more general form of mutation would allow a small and random number of simple moves to be applied to the thought.

$$\boxed{\textbf{mop9} \quad : \text{CROSSOVER: } thought1, thought2 \rightarrow thought_{crossover}}$$

This operation starts with two thoughts $thought1 = \sigma_1(g_{11}, g_{12}, \ldots, g_{1n_1})$, and $thought2 = \sigma_2(g_{21}, g_{22}, \ldots, g_{2n_2})$ and forms a new connector by combining a sub-connector $\sigma_1' \subset \sigma_1$ with a sub-connector $\sigma_2' \subset \sigma_2$. Keep generators as they are and form a new *thought* with the connector $\sigma_1' \cup \sigma_2'$. For example, see Figure 3.17.

Figure 3.17:

The reader will have noticed that we treat thoughts more or less as biological organisms. The crossover operation, in particular, is similar to what occurs in genetics. We shall also need the operator

$$\boxed{\mathbf{mop10} = \text{PERSONALITY CHANGE: } A \to A_{personality}}$$

that makes changes in the values of $A(self, \cdot)$ so that the MIND's behavior changes probabilistically.

3.8 Birth and Death of Thoughts

We certainly do not think of a mind as a static system, instead it will develop in time. As already mentioned, under free associations ideas and fragments of ideas will appear according to a probability measure $P = P(t)$ changing with time t but only slowly with time scales as minutes and days rather than seconds and milliseconds. In this view we could claim that what we are constructing is a theory of the *artificial life of thoughts*.

3.8.1 *Competition among Unconscious Thoughts*

Say that the current configuration *thought* $\in \mathcal{C}(\mathcal{R})$ has been decomposed into the top-thoughts

$$thought = top\_thought(g_1) \oplus top\_thought(g_2) \oplus \ldots; g_p \in tops. \quad (3.37)$$

Let us calculate the energies

$$E[top\_thought(g_k)] = -log[P\{top\_thought(g_k)\}]; \quad k = 1, 2, \ldots, p \quad (3.38)$$

determined by the current probability measure and its associated energetics $q(\cdot), a(\cdot, \cdot)$. Hence an energy can be found as

$$E_p = log\, n! - n log\, \mu + \sum_{i \in \sigma_p} q(g_i) + \sum_{(k,k') \in \sigma_p} a(g_i, g_{i'});$$

$$k = (i, j); k' = (i', j'). \quad (3.39)$$

In the random collection of sub-thoughts *they compete with each other for existence on a conscious level.* This may remind a reader of the role of the censor mechanism in Freudian psychoanalysis, but that is not our intention. Instead we consider the thinking process as a *struggle between unconscious thoughts in a thought chatter.* The competition is decided in terms of their energies, but it is not a deterministic decision process. Instead, their associated probabilities

$$\pi_p = exp[-E_p/T] \tag{3.40}$$

control the choice of the winning one, so that, on the average, low energies are favored.

For a hot mind, $T \gg 1$, the mind is a boiling cauldron of competing chaotic thoughts in the unconscious. Eventually the mind crystallizes into connected thoughts, reaching the level of conscious thought. For lower temperature the competing thought are less chaotic and the mind will settle down faster.

It is possible to study the energy relation that govern the mental processes of the MIND. Let us consider the reaction *thought1* $\rightarrow$ *thought2* with the associated energies E_1, E_2 and where the two thoughts have the representations *thought1* $= \sigma_1(g_{11}, g_{12}, g_{13}, \ldots, g_{1n_1})$ and *thought2* $= \sigma_2(g_{21}, g_{22}, g_{23}, \ldots, g_{2n_2})$. The energy difference is then

$$\Delta E = E_2 - E_1 = \sum_{i=1}^{n_2} \left[q(g_{2i}) + log\frac{i}{\mu} \right] + \sum_{(k,k')\in\sigma_2} a(g_{2i}, g_{2i'})$$

$$- \sum_{i=1}^{n_1} \left[q(g_{1i}) + log\frac{i}{\mu} \right] + \sum_{(k,k')\in\sigma_1} a(g_{1i}, g_{1i'}); \quad k = (i,j); k' = (i',j'),$$

$$\tag{3.41}$$

this is for $n_2 \geq n_1$; in the opposite case a minor modification is needed.

The energy equation simplifies in special cases of which we mention only a few. First, when *thought2* $= \sigma_3(thought1, g)$, where the connector σ_3 connects the new elementary idea g with $g_{1i_1}, g_{1i_2}, \ldots, g_{1i_s}; s = 1, 2,$ or 3. Then

$$\Delta E = q(g) + log\frac{n+1}{\mu} + \sum_{t=1}^{s} a(g, g_{1t}). \tag{3.42}$$

The choice of g that minimizes energy is then

$$g_{min} = arg\ min_g \left[q(g) + \sum_{t=1}^{s} a(g, g_{1t}) \right] \tag{3.43}$$

and this represents a tendency to a conditional ground state.

Second, if an elementary idea g_{1r} in *thought1* is deleted, and if it connects to other elementary ideas $g_{1i_1}, g_{1i_2}, \ldots$, then the energy difference is

$$\Delta E = -q(g_{1r}) - log\frac{r}{\mu} - \sum_t a(g_{1r}, g_{1t}).\tag{3.44}$$

Still another simple move, the elementary idea g_{1r} in *thought1* is changed to g and g_{1r} connects to $g_{1i_1}, g_{1i_2}, \ldots$, then the energy difference becomes

$$\Delta E = q(g) - q(g_{1r} + sum_t[a(g, g_{1t} - a(g, g_{1r})].\tag{3.45}$$

Compare with the discussion of randomization using MCMC in Section 3.5. The tendency to a ground state corresponds to the most likely reaction of MIND to a given situation.

As time goes on the energy changes and the resulting *energy development* is given by a function $E(t)$. An example is shown in Figure 3.18.

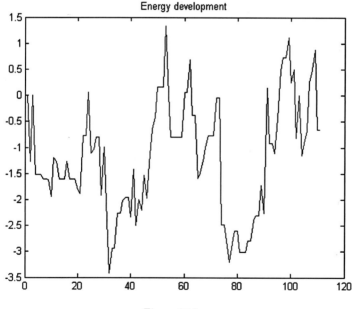

Figure 3.18:

The behavior seems to change at $t \approx 30$ and $t \approx 80$, probably due to a change in the mental soup in which MIND is immersed, in other words a change of theme (genre) that occurs at Markovian time points.

Part III
Two Personalities

Chapter 4

Building a GOLEM, a Thinking Machine

"But how can I make a Golem?"
thought Great Rabbi Loew

As described earlier we shall judge a mind model through the performance of a software realization of the model. We could say that we shall build a *GOLEM*, an artificial creature with some thinking abilities. A GOLEM could be said to belong to the sphere of *artificial life*.

But can we build a GOLEM using the principles announced above? That is, can we present a concrete system in the form of a computer program, that exhibits some of the characteristics we believe describe the human mind? We shall develop such a system in computational form, a first attempt, admittedly not very successful, but hopefully to be followed by a series of successively more sophisticated systems, perhaps culminating in one with a reasonably anthropoid behavior.

For the sake of programming ease, but at the cost of loss of speed of computation, we have selected MATLAB as the programming language.

4.1 Data Structures for the Mind

We believe that the choice of data structures is of more than peripheral interest. Indeed, the architecture of a GOLEM must be expressed in terms of data structures. The data structures we propose in the following are not arbitrary, but are the result of careful consideration, and likely to be the preferred choice in future realization of GOLEMs even if they are expressed in a different programming language and with more complex implementation.

The form of these structures has proven to be efficient. We recommend that the reader takes a careful look at the program code given in Appendix E.

4.1.1 *Data Structures for Elementary Ideas*

Ideas will have three attributes: name, level and modality. To handle this efficiently we shall let the generator space G be a MATLAB *structure* with the fields (1) name, as a character string, (2) level, as a numeric scalar, and (3) modality, also as a numeric scalar representing names in a variable "modalities". We enumerate G by an index g so that the gth one is

$$G(g) \in G; g = 1, 2, \ldots, r \qquad (4.1)$$

with three fields: the name $G(g).name$, the level $G(g).level$, and the modality $G(g).modality$.

To make the following concrete we shall use examples to clarify what we have in mind. The actual software that we shall use is going to be much more extensive but constructed in the same way as indicated by the examples. Some of the 1-level generators could be

$G(1) =$
name: 'man', level: 1 modality: 1
$G(2) =$
name: 'boy', level: 1 modality: 1
$G(3) =$
name: 'self', level: 1 modality: 1
$G(4) =$
name: 'Peter', level: 1 modality: 1
and some of other modalities:
$G(30) =$
name: 'chair', level: 1 modality: 8
$G(100) =$
name: 'jump', level: 2 modality: 28
$G(120) =$
name: 'today', level: 3 modality: 38

We could use for example the modalities (many more have been added in the MATLAB implementation)

1: humanM, M for male
2: humanF, F for female
3: animalM

   4: animalF
   5: food
   6: vehicle
   7: building
   8: furniture
   9: tool
10: machine
11: body part
12: idea transfer
13: apparel
14: capital
15: social group
16: size
17: color
18: smell
19: taste
20: sound
21: emotion
22: affect
23: hunger
24: greed
25: awareness
26: family relation
27: social relation
28: movement
29: eat
30: feel
31: likeHA, H for human, A for animal
32: likeT, T for things
33: activity
34: direction
35: quality
36: quantity
37: where
38: when
39: change hands
40: libidoH
41: libidoA
42: amicus relation

43: active ideas

44: new ideas

and many more. As we have noted before, signifiers like *man, likeT* and *changehands* should not be understood as words, but instead as concepts. We can get the modalities

$$humanM = \{man,\ boy,\ self,\ Peter,\ Paul, \ldots\} \qquad (4.2)$$

$$likeT = \{likeINAN,\ dislikeINAN, \ldots\} \qquad (4.3)$$

$$changehands = \{give,\ take, \ldots\}. \qquad (4.4)$$

The concept *humanM* means, for this mind, a man in general, a boy in general, the self = the carrier of this MIND, the particular man called Peter, or the particular man called Paul. The concept *LikeINAN* means to like or dislike something inanimate. The concept *changehands* means to give or to take, etc.

The connectivity of MIND will be given by the MATLAB *cell* "modtransfer" consisting of one cell for each modality, each cell with three subcells with numerical 3-vectors (possibly empty) as entries. For example cell no. 32: likeT in this MIND could look like

$$likeT = (1, 2; 5, 6, 7, 8; \emptyset), \qquad (4.5)$$

meaning that the modality is of arity 2 with the first bond extending downwards either to modality 1 or 2, the second bond to either 5, 6, 7, or 8 and no third bond. For simplicity we have limited the down-arity to three but that could easily be extended; we have not yet needed this. This ordering induces automatically a partial order in the generator space G.

4.1.2 *Data Structures for Thoughts*

To represent thoughts we use two arrays:

(1) an $n \times 2$ matrix "content" with $n =$ no. of generators

$$content = \begin{pmatrix} h_1 & g_1 \\ h_2 & g_2 \\ \ldots \ldots \\ h_n & g_n \end{pmatrix} \qquad (4.6)$$

where $(h_1, h_2, \ldots, h_n)$ means the *set* of generators in the configuration, expressed in h-coordinates and $(g_1, g_2, \ldots, g_n)$ the *multiset* of generators expressed in g-coordinates. The h's are assigned to generators as they appear one after another during the mental processes, numbering them consecutively, so that all the h's are distinct in contrast to the g's that can take the same values more than once; a thought can contain reference to, for example, "man" more than once.

(2) an $m \times 3$ matrix "connector", with $m =$ no. of connections

$$connector = \begin{pmatrix} j_{11} & j_{12} & j_{13} \\ j_{21} & j_{22} & j_{23} \\ \cdots & \cdots & \cdots \\ j_{m1} & j_{m2} & j_{m3} \end{pmatrix}. \tag{4.7}$$

This matrix has three columns for each row, i.e. connection. For the first segment j_{11} is the h-coordinate of the start of the downward segment, j_{12} is the h-coordinate of the end segment, and j_{13} is the j-coordinate of the generator from which the downward segment emanates, and so on for the other connections of this thought. See Figure 4.1.

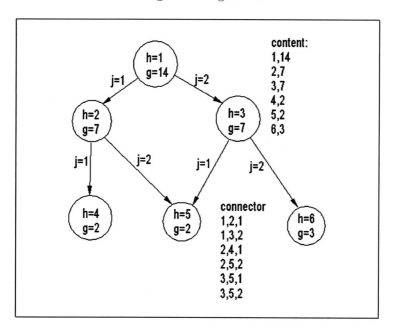

Figure 4.1:

We shall pay some attention to top-ideas of level 2 including at most 3 generators on level 1. Of course this reduces the intellectual power of the mind to the extent that it is unable to operate with abstractions on higher levels as far as top-ideas are concerned, but it can handle more complex abstractions by other means. We use the following data structures for such thoughts. If the top of a "thought" is $g_{top} = g_0$ and the subordinated generators are $g_1, \ldots, g_p$ expressed in g-coordinates, and with p at most equal to 3, we shall enumerate it with the *Goedel number*

$$goedel(thought) = \sum_{k=0}^{p} r^{g_k}; \quad r = |G|, \tag{4.8}$$

in other words, we use the base r radix representation.

4.1.3   *Energies of Thoughts and Genres*

It is easier to find suitable data structures for the mental energies. Indeed, we shall let q be a numeric r-vector and a be a numeric $r \times r$ matrix. The same data structures apply for the weight function $Q(g) = exp[-q(g)]; g = 1, 2, \ldots, r$ and the acceptor function (matrix) $A(g_1, g_2) = exp[-a(g_1, g_2)]; g_1, g_2 = 1, 2, \ldots, r$.

This makes it easy to represent genres (themes). Consider a genre called *genre* $\subset G$ consisting of the ideas that characterize this *genre*. Sometimes we modify the Q vector to take just two values: *max* and *min*

$$Q(g) = max; x \in genre; \quad Q(g) = min; g \notin genre. \tag{4.9}$$

Actually we shall use a somewhat more involved modification that will make it possible to account for the development of the mind including changes in genre energies.

As examples of the genres of the mind that we will use we mention the following:

(1) *emotional relation HA* between humans and animals

(2) *ownership* among humans and property

(3) *play pets* for human and pets

(4) *work* for humans

(5) *relax* for humans

(6) *movement* for humans and animals

(7) *interior design* for house and home

(8) *sports* for humans

(9) *reasoning* among humans, not purely logical but also, well, *unreasonable reasoning*

(10) *talking* among humans

(11) *eating* among humans and animals

(12) *objects* about inanimate objects

(13) *abstract thinking* with $Q = max$ for those g's for which $MOD(g) = g$

(14) *emotionalHH* about emotional relations between humans.

We shall also allow Boolean combinations of genres, for example *work* $\vee$ *objects*, meaning to work with some object, as well as more involved Boolean expressions.

4.1.4 *Composite Moves*

The data structure of a driver is a bit more complicated. It will consist of four parts:

(1) *change-thought* is an $2 \times n_{thought}$ Matlab cell; $n_{thought}$ is the size of the "sub-thought" that the mind is currently thinking about. For each subcell, $k = 1, 2, \ldots, n_{thought}$, a choice is made between (a) deleting the idea, or (b) keeping it unchanged, or (c) change to another g-value, or (d) choose a random a new g-value from a given set.

(2) *ad content* adds a set of new idea.

(3) *ad connector* adds connections but only inside the "sub-thought".

(4) *delete connector* deletes connections but only within the "sub-thought".

We have already seen a number of examples of drivers.

4.2 Program Hierarchy for the Mind

The GOLEM code is complicated and deserves the reader's attention: it includes many ideas and programming devices that have not been discussed in the text. Therefore *we recommend that a reader who wants to really understand the working of GOLEM to at least glance through the code given in Appendix E* and, in particular, read more carefully the main program *think*.

4.3 Putting It All Together

To build a GOLEM by successively introducing new entities we can proceed as follows (a) To introduce a new generator in an existing modality use *set_G*, followed by redefinition of the following MIND arrays:

gs_in_mods that finds ideas contained in a given modality followed by *get_levels*, *get_mod_transfer* giving the set of modalities for given modalities and the inverse mapping *get_mod_transfer_inv*,

set_Qs, *set_As* modifies the personality parameters A, Q one idea at a time,

set_g_mod defines modalities,

set_mod_omegas defines arities.

(b) To introduce a new modality use *set_modalities* followed by *get_levels*.

(c) Then use *print_G* to print the generator space with numbers and *print_modalities* to print modalities with names.

(d) Use *see_modality* to display a single modality graphically and *see_mind* to display the current configuration.

The above family of programs is combined into the *main function* "think" which displays a menu allowing the user to choose between the following alternatives:

(1) *ThinkingDrivenbyThemes*. This is the main mode of "think" with several options for the themes.

(2) *ContinuousThought*. In this mode the MIND trajectory jumps between different themes and creates new ideas occasionally.

(3) *ThinkingDrivenbyExternalInputsoftheMind*. The user inputs elementary ideas and the MIND makes *inference* from them to build a new thought.

(4) *FreeAssociations* where the trajectory through mind space consists of small steps of simple moves following the probability measure P, not driven by any other outer or inner forces. The result is fairly chaotic, unorganized thinking.

(5) *SetPersonalityProfile* in which the user defines a personality of "self".

(6) *SetMindLinkages* sets the mind parameters Q and A for a given personality profile.

(7) *TheVisibleMind* displays the connectivity of the MIND.

(8) *SeeCreatedIdeas* displays the new created ideas.

4.4 A GOLEM Alive?

Now let us see what sort of thought patterns are generated by the GOLEM anthropoid. The best way of studying the behavior of the program is of course to download the code and experiment with it oneself; the user is strongly encouraged to do this. Here we only present some snapshots and hope that they give at least some idea of the functioning of this MIND. Let us recall, however, that we do not view ideas and thoughts as words and sentences; instead we consider thinking as a flux of emotions, impressions, vague feelings, etc. The fact that the following diagrams involve words is just an admission that we do not (yet) have access to better representations than the verbal ones.

4.4.1 Free Associations

To begin with, let the GOLEM move freely through its mental space, not influenced by inner or outer constraints. Make the Q and A functions constant and so that the bindings are quite weak: one simple idea that has occurred to the MIND has little influence on the following ones. The partial ordering that we have imposed via the modality lattice prevents the resulting thoughts from being wildly meaningless, but the semantics is far from consistent; how to improve this will be seen later on.

As the program executes it shows a sequence of snapshots of the mind, one mind state is followed by another struggling to reach the level of consciousness. Here we can only show a few of the snapshots; executing the software gives a better idea of how the MIND is working in this mode. In Figures 4.2–4.5 we see some mind states under (very) free associations.

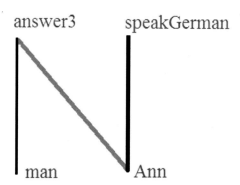

Figure 4.2:

In Figure 4.2 a man answers Ann who speaks German. The thought is incomplete; the arity of "answer3" is 3, but only two of its out-bonds are connected, so that it had not reached the level of consciousness.

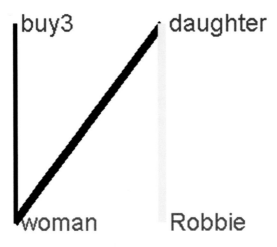

Figure 4.3:

Figure 4.3, a woman is the daughter of Robbie, but what does she buy and from whom? An incomplete thought, $\omega(buy3) = 3$.

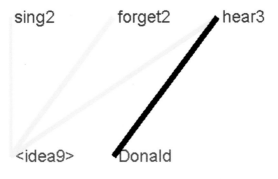

Figure 4.4:

In Figure 4.4 Donald hears an idea, but who sings and who forgets? The meaning is not clear due to the incompleteness of the thought!

Figure 4.5, Peter strokes the puppy who whimpers — finally a complete thought.

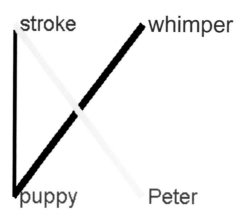

Figure 4.5:

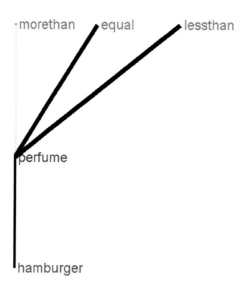

Figure 4.6:

In Figure 4.6, the thinking is disorganized, perhaps the GOLEM is dreaming about the smell of a hamburger. The ideas on the third level seem unrelated, actually inconsistent. However, the user can instruct the GOLEM to concentrate its thinking, *try to connect sub-thoughts* that appeared disjoint and independent. The way to do this is to choose the option "Concentrated Thought". The resulting idea will appear concentrated with its sub-ideas connected to the extent that regularity and the structure formula allow. This option can be applied in some other modes of thinking too. It will have a noticeable effect only when the original connector is not fully closed.

4.4.2 Inferential Thinking

Now we force the GOLEM to start from given external inputs and continue it further by the inference process described in Section 3.7.6. Say that GOLEM starts with the MIND's input being "cash", *genre* = BUSINESS, the one-idea thought in Figure 4.7 with the inference in Figure 4.8: a visitor gives cash to Carin or with the input "aspirin" an inference is shown in Figure 4.9 where Bob swallows the aspirin but with some additional thought

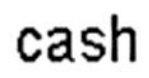

cash

Figure 4.7:

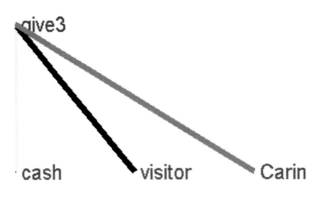

Figure 4.8:

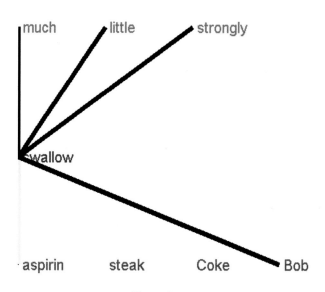

Figure 4.9:

chatter; note the inconsistency on level 3 which is to be expected in a thought chatter. Such imperfections actually add to the verisimilitude of GOLEM.

Starting with the idea of "Republican" the inference is in Figure 4.10.

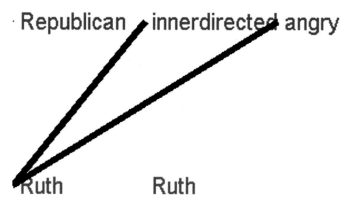

Figure 4.10:

This is incomplete and more or less meaningless, free associations can lead to nonsensical thoughts. But human thought can develop in strange ways!

4.4.3 *Associations Driven by Themes*

GOLEM can carry out thematic thinking (genres). Once the inputs are defined, GOLEM can start thinking, influenced by the inputs. Here is one thought from the theme Sports with Linda playing in Figure 4.11.

Linda plays dice with a boy. She also turns and hikes badly. Well, barely understandable? Another thematic thought from the theme Business.

In Figure 4.12, Donald carries out complicated transactions with belongings changing hands. GOLEM had not yet settled down to a conscious state, note that $\omega(sell3) = 3$, but "*sell3*" has only two connected out-bonds. For the theme Pets we get Figure 4.13.

The thought is highly incomplete. The only completed sub-thought is that Rufsan is brown, but it is not clear who whistles at her and tells her she is a bad dog (repeatedly). We believe that such incompleteness is typical for some human thinking. And the theme Business again, in Figure 4.14.

Eve buys a lot; a complete thought.

In these figures we have not shown the thought chatter that induced the resulting thought; that can be seen by running the software and is quite instructive.

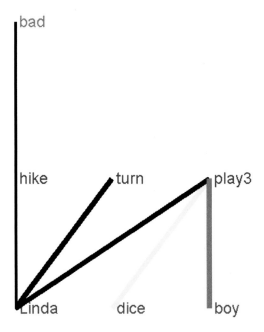

Figure 4.11:

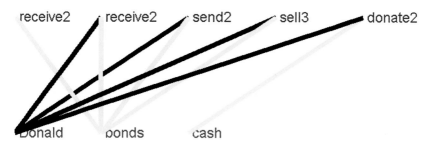

Figure 4.12:

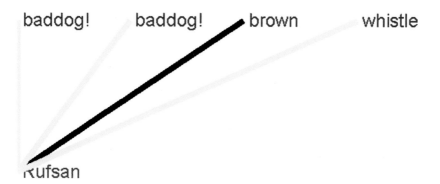

Figure 4.13:

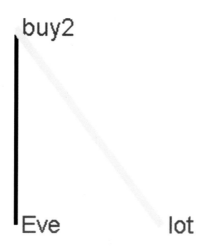

Figure 4.14:

4.4.4 *Continuous Thought*

This is an important option and deserves our attention. Among all the sub-thoughts, complete or incomplete, that exist in the mind at any given mind, only some reach the level of consciousness as was discussed earlier. To see how this happens, execute option "Continuous Thinking" that shows thought chatter and later the resulting thought. It moves via a Markov chain through the themes. The user is asked for the duration of thinking, choose a low number. During the thinking the direction of the mind trajectory may change, if this happens it is announced on the screen. Also, if a new idea is created and added to the generator space it is announced. New ideas can be displayed using the option "See New Created Ideas" in GOLEM. For example, Figure 4.15 in which Lisbeth tells Spot he is a bad dog and also pinches Rusty who turns. Lisbeth is tanned brown.

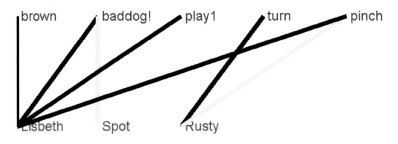

Figure 4.15:

A thought chatter, actually a completed thought, is shown in Figure 4.16, where the visitor is smiling while buying. Or, in Figure 4.17,

Figure 4.16:

94

with no resulting thought, the mind is at rest!

Figure 4.17:

Again continuous thinking: Spot is jumping in Figure 4.18.

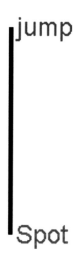

Figure 4.18:

In Figure 4.19 Helen strokes Bob who plays, a complete thought.

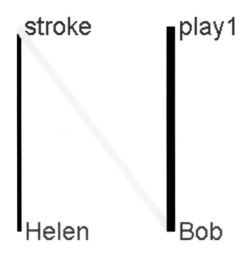

Figure 4.19:

4.4.5 *See Created Ideas*

To display ideas that have been created by GOLEM and added to the generator space choose the option "See Created Ideas". For example:

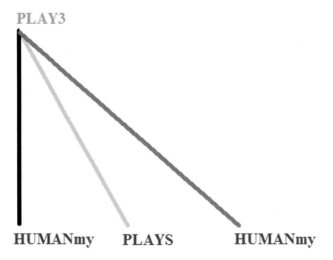

Figure 4.20:

Two young males play unspecified "plays" with each other in Figure 4.20.

4.5 Drivers

We have only experienced with a few drivers. One of them is *love_driver_1*; in Matlab form as a "cell(6,1)" with the first sub-cell

$$
\begin{pmatrix}
change & 247 \\
same & [] \\
same & []
\end{pmatrix},
$$

the three next sub-cells empty (no generators or connections added), the fourth one .8 (activation probability, and the sixth one the domain of the driver (246, humanM, humanF). This driver searches the configuration for top-2ideas that belong to the driver. If it finds one, it replaces generator $g = 246$, meaning "love", with generator $= 247$, meaning "desire". We use the program "build-driver" for constructing drivers and "execute-driver" for executing them. We get for example starting with the idea "Jim love Joanie" in Figure 4.21, driven into the new idea "Jim desire Joanie" in Figure 4.22.

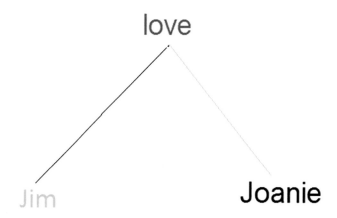

Figure 4.21:

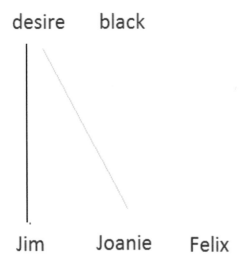

desire   black

Jim   Joanie   Felix

Figure 4.22:

4.5.1 *Generalizing Top-Ideas*

One of the options for GOLEM is to determine the top-2ideas currently in consciousness, and then generalize them (first order) into the modality lattice to get a thought pattern. We get for example Figure 4.23 signifying the concept of a moving young male. And the thought in Figure 4.24, which shows the thought pattern when a capital transactions involving jewelry takes place to a female adult.

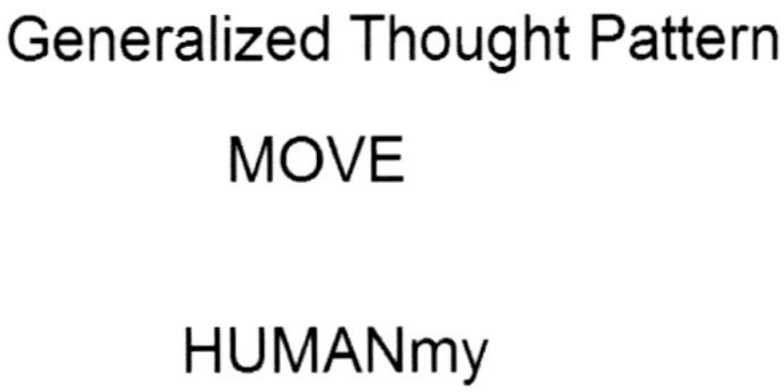

Generalized Thought Pattern

MOVE

HUMANmy

Figure 4.23:

Genralized Thought Pattern

COMMERCIAL2
HUMANfa   JEWELRY

Figure 4.24:

4.6   Judging the Behavior of GOLEM

> *When the immortal Dr. Johnson had heard a woman preach*
> *he said it reminded him of a dog walking on its hind legs,*
> *it was not well done but it was remarkable that it could*
> *be done at all.*
> (S. Johnson, 1709–1784)

How well does GOLEM imitate human thinking? The code is working but it does not work well. It clearly attempts to do so but with mixed results. Under Free Associations the thinking ought to be chaotic but GOLEM's thoughts appear *very* chaotic. One is led to apply Dr. Johnson's evaluation. The connections between sub-thoughts are *too* random, they should be better controlled by the probability measure used. The performance is better under Continuous Thought and Thinking Driven by Themes, and this gives a hint for improvement. The set of themes ought to be refined into many more and more specific, narrower, ones. As one theme is followed by another the direction of the GOLEM trajectory changes, but in between jumps the probabilistic dependence seems adequate.

To improve the semantics the generator space must also be extended. In the current version we have used $r = 726$ generators organized into $M = 180$ modalities. This is clearly insufficient. Perhaps $r = 5000-10000$ and $M \approx 1000$ would be adequate. To implement this would require more manpower than what the author has had available. It should be mentioned, however, that a substantial research effort in AI has been directed to defining a large set of concepts and relations between concepts; see for example www.opencyc.org. Perhaps this could be used to extend GOLEM. Also, the modalities should take into account a *taxonomy of ideas*, expressing how human knowledge can be organized into fine categories. This will require more levels representing different degrees of abstraction.

Perhaps GOLEM should also produce outputs: movement, speech, external reactions, limbic response and so on. We do not see how this can be attained and how to express such outputs. Possibly by using avatars. This will be neccessary to allow for interactions between GOLEMs to be discussed below.

Although GOLEM's performance in imitating the human mind is not impressive, it indicates that a degree of verisimilitude can be achieved by a probabilistic algorithm. When de La Mettrie opened a discussion on the theme *L'Homme machine* it began a discourse that would have delighted the School Men. We shall certainly avoid getting involved in this morass of vague philosophizing. Instead of the metaphor of a machine, with its image of cog wheels and levers, or transistors on silicon, we shall only claim that *the mind can be viewed as an entity that is subject to laws, probabilistic to be sure, but nevertheless regulated by definite rules.* Our main task is therefore to formulate and verify/falsify hypothetical laws of the mind.

In spite of the limited success of GOLEM our temporary conclusion is: *The human mind can be understood.*

4.6.1 *Analysis of a Virtual MIND*

Say that we observe the output of a virtual MIND without knowing its inner workings, and that we want to understand it. Here the term "understand" means knowing, at least partly, the parameters that characterize the mind: $G, \mathcal{M}, Q, A$ and possibly others. One could say that we want to perform *psychoanalysis without Freud*. It is known in general pattern theory how to estimate, e.g., the acceptor function A. See GPT Chapter 20 and also Besag (1974), Osborn (1986), where however the connector graph σ is supposed to be fixed and not random as in GOLEM.

It will be more appealing to the intuition to use other parameters for the analysis. Indeed, Q and A do not contain probabilities as elements as may have been thought at first glance. For example, the entries in the Q-vector can be greater than one. Q and A are needed for the probabilistic generation of thoughts but are not simply related to probabilities of simple events. Instead we shall introduce parameters that have a direct interpretation but are not simply related to the Q and A. This is strictly tentative.

For any positive content size n and any generator $g \in G$, consider the average of the conditional probabilities

$$f(g|n) = \frac{1}{|\sigma|} \sum_{i=1}^{n} P(g_i = g : |\sigma| = n) \tag{4.10}$$

and

$$f(g) = \sum_{n=1}^{\infty} p(n)f(g|n) \tag{4.11}$$

so that $f(g)$ measures the possibility of MIND making use of the elementary idea g. Further, the expression

$$F(genre) = \frac{1}{|genre|} \sum_{g \in genre \subset GENRE} f(g) \tag{4.12}$$

measures the *propensity of a particular genre*.

Then we can estimate these parameters in a straightforward way. We simply replace the probabilities $P(g_i = g : |\sigma| = n)$ and $p(n)$ by the respective observed relative frequencies. But we can reach deeper into the structure of MIND. Indeed, let us fix two thought patterns $PATTERN \in \mathcal{P}$ and $PATTERN'$, and consider two (random) consecutive thoughts, $thought(t)$ and $thought(t+1)$ occurring to MIND at time points t and $t+1$. Introduce the conditional probability

$$Prob = P\{PATTERN' \in thought(t+1)|PATTERN \in thought(t)\} \tag{4.13}$$

measuring the likelihood that $PATTERN$ is followed by $PATTERN'$. We do not insist on any cause-effect relation, just temporal sequentiality.

For example, if $PATTERN$ is a pattern representing one person, the self, challenging another, and $PATTERN'$ represents violent action, then $Prob$ is a mind parameter with a rather clear interpretation as aggressiveness. Or, if $PATTERN$ stands for self and $PATTERN'$ for sadness, then $Prob$ could be understood as a tendency to depression.

It should be remarked that $PATTERN'$ corresponds to a sub-graph with many inputs, this can imply that this pattern is likely to be activated. This statement should be qualified by pointing out that the likelihood depends upon how the A-values for these in-bonds have been modified by MIND's experiences during its development.

Where Do We Go From Here? In spite of its less than impressive performance the GOLEM points the way to the development of more powerful artificial minds. The improvements suggested in the previous section will require much work, in particular the development of auxiliary programs (see below), but nothing new in principle. However, we have started to see some more challenging extensions.

The notion of driver discussed above seems essential. We defined just a few drivers but could easily add to them in the spirit of the composite moves in Section 4.5 using the program "build-driver". But this does not seem the right way to go. Instead the creation of new drives ought to be wholly or partly automated, maybe through energy based extremum principles. As GOLEM is experiencing new inputs from the external world, and perhaps from interactions from other individuals, it ought to solidify its experiences into drivers. This should happen over long intervals of time. It is not yet clear how to arrange this.

The GOLEM should live in a world inhabited by other GOLEMs, similar but not identical to it. They should exchange ideas and modify themselves as a result of such symbiosis — *a mind game*. For this it is necessary that all the GOLEMs have their out-inputs in the same format: compatibility.

Once in- and output are defined it seems natural to *analyze the mind in terms of conventional personality types*; we have used some crude types in the program *think*. See Brand (2002) for a catalog of personality categorizations suggested in the psychological literature.

Earlier we discussed the decisive role of randomness in the study of human thinking. Actually, a more radical approach would be to think of ideas as *clouds of uncertainties* described by probability densities in a high dimensional feature space. The calculus of ideas that we have proposed would then operate on probability densities, a bit similar to the role of wave functions in quantum mechanics. At the moment it is far from clear how to make this precise; some adventurous colleague may be tempted to look more closely into this possibility.

4.7 Not Yet Implemented

The following additions to GOLEM seem natural but have not yet been implemented.

(1) One should allow a generator in a thought to be dominated by at most one generator for each modality. This is to avoid thoughts like

(*small, big, house*). An earlier version of GOLEM had this constraint realized but was later excluded.

(2) The mind operations MUTATE, SPECIALIZE and CROSSOVER have not been included in the code. The two first ones can easily be implemented with minor changes in the existing code, but CROSSOVER would require some effort.

(3) GOLEM does not (at present) delete new ideas when they are not reinforced by repeated occurrence. They should be deleted if ideas with the same *content* are not replicated often enough.

(4) GOLEM can perform *link analysis.* For a given set of ideas (concepts) running the GOLEM as an interpolator will discover links and attach weights to them if desired. This could be of considerable practical use, to "connect the dots" to use a standard cliché.

(5) The thinking simulated by GOLEM is fairly slow, in particular if the speed of the computer is less than 2 GHz. If one had access to parallel hardware it should be possible to achieve much better speed if each level in the configuration for building "thought" was treated at once. May we suggest that this is reminiscent to the columnar organization of the brain?

(6) In Section 3.1 we mentioned the possibility of proving probabilistic limit theorems for the construction of approximately optimal critical regions designed for testing potential abnormality. This has not been done but could also be useful for the analytical understanding of thought patterns created by GOLEM-like creatures.

4.8   LEGACY

We have introduced some mind algebras whose main role was to illustrate the general concepts of thought patterns by concrete examples. The result was not very convincing in terms of realism and anthropomorphic behavior. Now we shall be more ambitious: the choice of mind parameters $G, Q, A, \ldots$ will be made more carefully. Indeed, we shall try to represent a *particular* mind. But which mind shall we select? Obviously the one best known to the author is his own. A disadvantage is that readers who do not know the author familiarly will find it hard to interpret some of the thoughts. Admitting that the knowledge available through introspection is completely subjective we shall rely on it to select the mind parameters.

However, we are conscious of widespread suspicion of introspection as a tool for studying the mind. For example, one of the giants in the history of mind studies, Francis Crick, proposes a research attitude: "Concentrate on the processes in the brain that are most directly responsible for consciousness", and his is the dominating positivist view among serious researchers. All modern science is based on experimental observation leading to testable hypothesis and being able to falsify them. What we are doing is less orthodox; to get at least some support from the giants among psychologists we refer to William James: "Introspective observation is what we have to rely on first and foremost and always."

Anyway, with less than impressive support from the authorities we shall go ahead intrepidly and try to select mind parameters from our own thinking. This requires a lot of work, it is very time consuming. Indeed, we have to choose thousands of generators, not to mention the Q and A parameters. They have to fit the mind we are trying to represent and this will require a good deal of thought. The parameters will express the environments in which the mind lives, both material and mental. Also the intellectual and emotional habits of the particular mind, friendships and families, work milieu and hobby activities. Altogether an impressive endeavor, see the next section.

Once this has been done in a satisfactory way we run the LEGACY software with specified parameter values. The software can be found on the Internet in http://people.sissa.it/~pirmorad/patterns_of_thought.html.

This will serve as a memory, a legacy, of this mind to its remaining family members. This is like an autobiography but with the major difference that it does not simply enumerate memories of persons, things, events, ... It also shows the mind in action, how it reacts and associates, creates new ideas, remembers and forgets and so on. It is a thinking memory: a reactive agent. Then it is another question how well we can make GOLEM represent the real mind. Here we only offer preliminary attempts but hope that other researchers will extend and complete the attempt as well as to write more sophisticated software.

4.9 Assembling Mental Pattern Parameters

To organize the selection of the generator space G of elementary ideas we shall use a formalized version of the procedure described in Section 5. With the decomposition in terms of levels

$$G = \cup_l G^l \tag{4.14}$$

we shall construct the subspaces G^l recursively. Assume that $G^1, G^2, \ldots$ have been constructed. For a given finite sequence of subsets $G^l_k(j); j = 1, 2, \ldots, \omega$, we shall introduce a finite number of new generators $g^l_k(n); n = 1, 2, 3, \ldots$ belonging to G^{l+1} to be created. Note that this construction is completely abstract with no reference to the properties of the mind. That is done instead by the constructor who will choose the $g^l_k(n)$ so that they correspond to characteristics of the mind in terms of the meaning of the already chosen sequence $G^l_k(\omega); \omega = 1, 2, \ldots$. The elements of this sequence will be the out-bonds of $g^l_k(n); n = 1, 2, 3, \ldots$ defining a modality. Note that this induces a modality structure due to the way a new generator $g^l_k(n)$ relates to the subset G^l_k.

As an example, let $G^l_k(1) =' humans', G^l_k(2) =' objects', G^l_k(3) =' humans'$ we could let $g^l_k =' give - to3$ and another generator $g^l_{l'}k =' get - from3$ or $G^l_l(1) =' human'$. $G^l_k(2) =' human'$ defining a new generator $g^l_k =' love2'$.

But how do we start the recursion, choosing G^1? In contrast to the above abstract procedure we will now make concrete assumptions about the meaning of this sub-space. Start with the partition

$$G^1 = G^{material} \cup G^{immaterial} \tag{4.15}$$

followed by

$$G^{material} = G^{animate} \cup G^{inanimate} \tag{4.16}$$

and perhaps

$$G^{animate} = G^{human} \cup G^{animal} \cup G^{flora} \tag{4.17}$$

and

$$G^{immaterial} = G^{active} \cup G^{passive}. \tag{4.18}$$

In the first example above we could let $G^l_k(1) = \{man, boy, girl, \ldots Ann, \ldots\}$ and $G^l_k(2) = \{book, chocolate, \ldots, computer, \ldots\}$.

Then we have to fill in generators in these sub-spaces, somewhat arbitrarily, and also attribute values to the arrays Q and A. The latter we have done simply by making all the entries of the arrays equal to 1. Note however that when GOLEM is running, things change. New ideas are created and added to G. Also the entries in Q and A are updated as described in Section 5.1; the mind is not staying the same but develops dynamically as influenced by the thought trajectory.

We have started LEGACY with about 1300 elementary ideas, but during its lifetime new (complex) ideas are generated as determined by the random occurrence of other ideas earlier.

Also, we started LEGACY with 327 modalities. One particular modality, *COMPLEX*, will hold ideas with more than one generator that have been created during the lifetime of the mind. For the moment we allow only ideas with $n = |content| < 5$.

We then divide G into themes and have used the following 12:

1. YOUTH
2. MIDDLE-AGE
3. OLD-AGE
4. WORK
5. SUMMER
6. FREE-TIME
7. ART
8. FEELINGS
9. FREE-ASSOCIATIONS
10. HOME
11. BUSINESS
12. REMEMBRANCE

The names are self-explanatory except for the last one *REMEM-BRANCE* which is intended to represent a meditative mind state when the ideas in COMPLEX are recalled and displayed.

The elementary ideas are arranged in four levels in the generator space G. In the first level we place the *objects*, i.e. ideas that make sense in isolation. This expresses a Wittgensteinian mode: his facts are the objects. On the second level we place ideas, *modifiers* that refer to objects and the third level will contain references to the modifiers, the *moderators*. Finally the fourth level will have the *existence ideas* like "true", "possible", "improbable". Or, schematically:

Level 4: existential

Level 3: moderating

Level 2: modifying

Level 1: objects.

This could lead to, for example, the thought train: *Ulf — happy — very — possible*.

As time goes on the mind creates new ideas and stores them in memory. This is done by encapsulation and the resulting ideas are put in the modality COMPLEX. For simplicity we have allowed at most $n = 4$ elementary ideas.

Note, however, that encapsulation can be iterated so that large complex ideas can be created.

To help create G we use special software to do the array processing, but even with such help the procedure is tiresome and time consuming.

In addition we create a MATLAB structure PIX with pictures and associate them to appropriate generators (ideas) or configurations. One could also introduce audio files corresponding to some ideas, perhaps utterances or sentences, but this has not been done in LEGACY.

4.10 Running the LEGACY

To run the software execute the command *legacy*. It takes a while when the computer is preparing the program; then a germ of thought is shown — this thought is unorganized. Then thought chatter takes over, picking a mental theme and tries to organize the thought into a conscious thought. When this has been achieved the result is shown as a dominating thought rather than thought chatter. Sometimes a complex idea is referred to and the user can ask for it to be resolved into elementary ideas and/or modalities representing the abstraction.

Occasionally an association to an image is realized and shown in the desk top. Less often a new idea is created and is shown together with its Goedel number. Observe the huge values of the Goedel numbers for the created ideas. This is in spite of the fact that we have only used top-ideas of height at most equal to 2; otherwise the Goedel numbers would be even larger. This indicates that a human mind forms only a thin slice through the universe of possible minds. Individuals are individualistic.

Sometimes it can happen that the Q and A arrays are updated: memory is modified due to the thought trajectory that has been experienced by the mind.

Now a few examples. They often refer to persons, places, things from the author's experience, as they should, and may not be known to the reader. In Figure 4.25 the thought indicates thirst, while the next one, Figure 4.26 says that *Stuart* is eating.

Figure 4.27 means that *Barbro*, a friend of the author, is reading *Stagnelius*, a Swedish poet.

In Figure 4.28 we "see" the thought that *Basilis* and *Don* discuss the grid model, while the next figure indicates that *Anders* and *Nik* speak Swedish to each other.

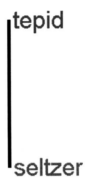

Figure 4.25:

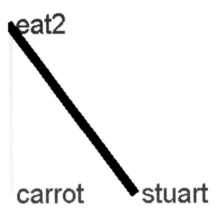

Figure 4.26:

Figure 4.27:

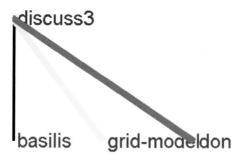

Figure 4.28:

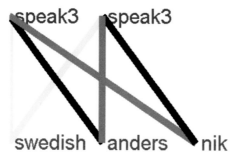

Figure 4.29:

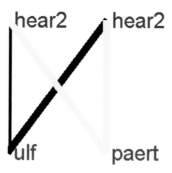

Figure 4.30:

Figure 4.30 says that *Ulf* listens to *Paert*, an Estonian composer. It is not clear how to interpret the double occurrence of *hear2*.

Figure 4.31:

Next, Figure 4.31 means the thought train "it is true that Marika now is preparing herring".

Figure 4.32 says that *idea*30 has been forgotten, where *idea*30 means that "Ulf listens to the piano", shown in Figure 4.33.

Figure 4.32:

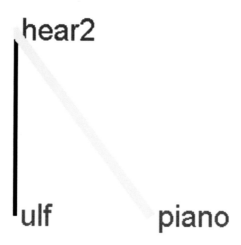

Figure 4.33:

After having run LEGACY a large number of times it is clear that it performs better than GOLEM. It is only occasionally that it produces thoughts that seem strange or at least irrelevant. Perhaps the modality structure of G should be made finer. However, on the whole we have achieved what we set out to do and look forward to further improvements.

To run LEGACY press "run legacy". Then it will wait for quite a while (depending on the speed of the computer) preparing for execution. Then a menu will appear — type in answers and so on.

Chapter 5

As Thinking Develops

5.1 Changes in Personality Parameter

As time goes on, the mind is evolving as the result of ideas that have been created and others forgotten. The *long term memory* is represented by the Q and A functions as well as by the evolving generator space G. If a generator g has occurred the effect will be assumed to be updated as

$$Q(g) \to remember_Q \times Q(g); remember_Q > 1 \qquad (5.1)$$

where the constant $remember_Q$ expresses the strengthening of memory concerning g. Each time that g does not occur the effect is

$$Q(g) \to forget_Q \times Q(g); forget_Q < 1 \qquad (5.2)$$

with another constant $forget_Q$ for the loss of memory, with $forget_Q$ closer to 1 than $remember_Q$. The acceptor function is modified in a similar way.

Hence we have the MEMORY operation

$$MEMORY : (Q, A) \mapsto (Q_{modified}, A_{modified}). \qquad (5.3)$$

When a new thought *idea* is added to G its Q-value is set proportional to the power $2^{iter(idea)}$ initially and will of course be modified later on due to new experiences and thinking.

It will sometimes happen that some newly created ideas coincide. To avoid misuse of memory we shall remove the copies. Actually, we shall do this as soon as the *content*'s are the same whether the *connector*'s are the same or not; recall that *content* is a multi-set. This is done for no other reason than to reduce thinking effort; isomorphism for graphs is a tricky business. Two ideas $idea1$ and $idea2$ will be considered different iff $content(idea1) \neq content(idea2)$. Periodically the memory will be updated by replacing two or more equal ideas by a single one: $\{idea1, idea2, \ldots, idea_k\} \rightarrow idea1$, removing its copies and setting $Q(idea1) = \sum_1^\nu Q(idea_\nu)$.

In other words, the ideas behave as organisms: they get born, they grow, compete and change, they die, and the population of ideas in G evolves over time. *The MIND has a life of its own.*

But what happens if the MIND is not exposed to any inputs, it just lives an isolated life? The statement in Appendix D answers this question. The MIND will degenerate more and more, limiting itself to a small subset of elementary ideas, namely those that were favored by the Q-vector at the very beginning of isolation.

5.2 Development of a Young Mind

A young mind, say that of an infant, starts out as a simple organism. As it grows more complex structures will appear as the result of sensory inputs. But how should the inputs from the senses be connected to the sites of the MIND? Let us think of an infant in the first stage of development in Piaget's stage: the sensorimotor stage. The child is learning objects as existing units outside the child itself. For example the concept *apple*. Perhaps something like the template in Figure 5.1.

The indicated connections are strong, but there may also be somewhat weaker ones connecting the template to, for example, location inputs. They should be weaker since their values are more variable and not occurring often enough to be frozen into long term memory. There may also be weak links to *red* and *yellow*; both values can occur, alone or even together.

For a later Piaget stage, when the child does not just identify objects, but also actions on objects, we may have a template as the one in Figure 5.2 for the concept *eat*, consisting of three elementary ideas: muscle movement for chewing, muscle movement for swallowing, and the sound of chewing.

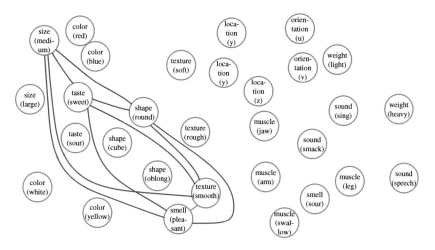

Figure 5.1:

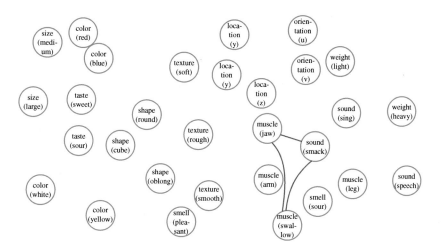

Figure 5.2:

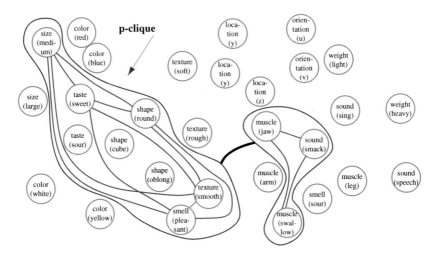

Figure 5.3:

With this done, the child's mind is ready for composite thought as in Figure 5.3, where the heavy black curve stands for a bus connecting the sites in the two templates.

5.3 GOLEM Evolves

After running GOLEM or LEGACY for a long time the MIND has changed: its linkage structure has been modified due to internal and external activities. To illustrate this look at Figure 5.4, that exhibits the linkages at an early stage of development, and Figure 5.5, where we see many more links established a long time later. Note in particular the increased activity close to the elementary idea "self", indicated by a small red star to the right in the diagram.

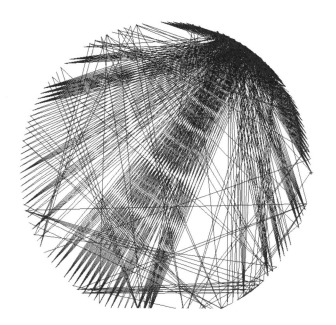

Figure 5.4:

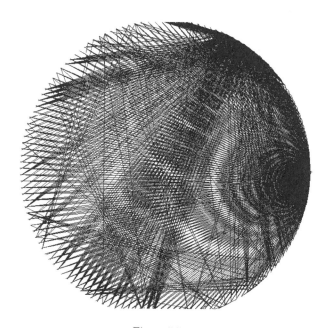

Figure 5.5:

This inspires to more experiments studying the mental development of MIND under different external environments and themes. How does the linkage structure change if GOLEM is run without external inputs? Or, if it is exposed to a single theme. And, if "self" has become very aggressive — what sort of inputs should one apply to MIND in order to improve the behavior: another option THERAPY? Much remains to be explored here.

Part IV
Realizing a Mind

Chapter 6

MIND and Brain

6.1 Connect to the Brain?

So far we have avoided any reference to a neural substrate for thought, to wit, the brain. But since we have already started down the slippery slope of speculation, we can just as well continue with some unbaked ideas of how to relate GOLEM and LEGACY to actual human thinking. Let us imagine the following experiment aimed at finding relations between MIND and the brain.

Using fMRI, say that we equip a patient in the magnet with special glasses for visual inputs and with ear phones for auditory inputs. The sensory inputs should be chosen so that they can be represented as "thoughts" in GOLEM. We then obtain a series of low resolution scans $I^{\mathcal{D}} = \{I^{\mathcal{D}}(1), I^{\mathcal{D}}(2), \ldots, I^{\mathcal{D}}(T)\}$ for the sensory inputs $thought(1), thought(2), \ldots, thought(T)$. Using deformable template techniques, see Grenander (1993), it may be possible to relate the observed blobs that have lighted up in the images to the various components of the brain. This will give us mappings

$$I^{\mathcal{D}}(t) \to \gamma(t) \tag{6.1}$$

with the γ's representing collections of brain components; $\gamma(t) \in \Gamma$.

Then we are confronted with a statistical estimation problem of general regression type: Find approximate relations

$$thought(t) \approx \gamma(t)). \tag{6.2}$$

To find such relations construct, for each t and i an arrow

$$g_i(t) \to \gamma(t) \tag{6.3}$$

for

$$thought(t) = \sigma(t)(g_1(t), g_2(t), \ldots, g_i(t), \ldots) \qquad (6.4)$$

one arrow for each brain component in $\gamma(t)$. This results in a *statistical map mind → brain*. This map tells us how primitive ideas are related to activities in the various brain components, and if we find broad channels in it, we have established a MIND/brain relation.

Can this experiment actually be carried out? We leave that question to researchers more familiar with brain research than the author.

We offer a simple example. The construction follows the general principles of Pattern Theory; see Grenander (1993), GPT, in particular Part I and Chapter 7. In order to make the discussion concrete we shall argue by special examples. What is lost in generality is gained in clarity. There is a danger in general philosophizing, difficulties may be hidden in vague propositions, but can be brought to the surface by limiting the discussion to special cases, the devil is in the details. Therefore, specify the details by a definite construction. To quote Carver Mead: "... you understand something when you can build it".

6.1.1 *Simple Ideas as Building Blocks*

Our starting point in this high level construction are the simple ideas, just as in the MIND, and we shall treat them as we did in Chapter 2, but with some modifications. The simple ideas range from very concrete concepts related to sensory inputs to abstractions successively built from more concrete ones. We shall denote simple ideas by $g, g_1, g_i, \ldots, idea_1, idea_2$ and so on, together forming an idea space G. They will be arranged in levels $l = 1, 2, 3, \ldots, L$, where L could be, for example, 6. All levels contains ideas connected to sensory inputs via some processing units. By senses we mean not only the classical five: vision, audition, smell, touch and taste, but also sensations due to hormonal and other body functions such as affects, feelings, hunger, muscular activity, ...; this is following Damasio (1999). Hence, ideas are not necessarily represented by words and sentences in a natural language, so that our approach is extra-linguistic. Thinking comes before language!

We will be guided by David Hume's radical proposition: "Though our thought seems to possess this unbounded liberty, we shall find, upon a nearer examination, that it is really confined within very narrow limits, and that all this creative power of the mind amounts to no more than the faculty of compounding, transposing, augmenting, or diminishing the materials afforded us by the senses and experience," a statement that is still valid.

We suggest the following temporary definition: *An idea is a set of nodes in the network together with is connections.*

To mention just a few simple ideas, we present the collection in Figure 6.1 with obvious interpretations, so that $idea_1 = $ 'soft', $idea_2 = $ 'coarse', ...

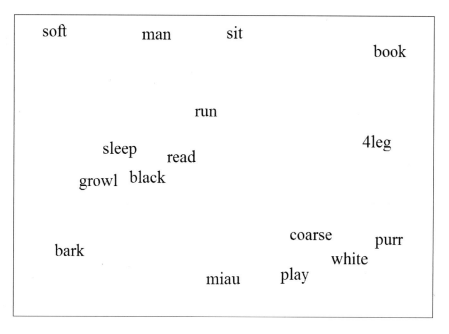

Figure 6.1:

6.1.2 *Connections between Simple Ideas*

Our construction is architectonic, following Immanuel Kant, with inter-level connections between adjacent levels $l, l + 1$, both ascending and descending.

CONCLUSION: *Association is all, no special metaphysical construct is needed.*

The intra-level connections are supposed to be locally dense, by which we mean that inside a given level the percentage of connections for a given distance between ideas decreases from 100 to zero as in Figure 6.2.

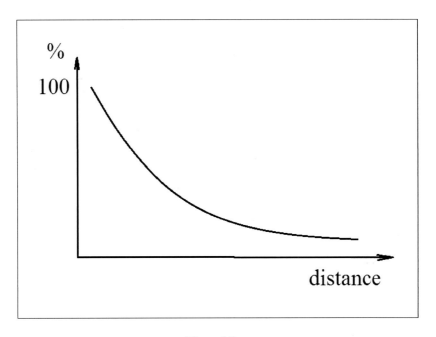

Figure 6.2:

This implies that any simple idea is connected, weakly or strongly, to all or most ideas close by. To organize simple ideas into thoughts we appeal to GPT and shall use the concept of *deformable template*, see Grenander (1993), Chapter 16. In the present context, an *idea template* will be a set of sites in the network and the deformation mechanism will consist of the choice of a subset of those sites. Such a deformation destroys information and the role of the mind is to try to recover it using knowledge stored in memory. In Figure 6.3 we see idea templates for "dog", "cat" and "read" as well as a small network; strong connections are shown as thick lines and weak ones are indicated by thin lines in the bottom right panel.

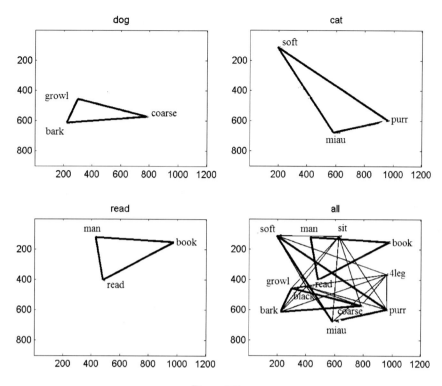

Figure 6.3:

6.1.3 Thoughts are Formed from Simple Ideas

The *acts of thinking* are expressed in the form of *thoughts*, combinations of simple ideas,

$$thought = \sigma(idea_1, idea_2, \ldots, idea_n) \qquad (6.5)$$

where σ stands for the graph connecting some of the simple ideas. If a thought has occurred repeatedly it may be *conserved*, memorized, as a template made up of simple idea templates. An $idea_i$ is located at site i and has a name, e.g. "red", "location x", "shape y", and can take the values *on* and *off*.

In the literature there is a concept *motif* which is similar to that of our template; see Sporns and Kotter (2004), but without any probabilistic super structure. See also Mumford (1992), first paragraph of Section 4.

6.1.4   *Weight and Acceptor Functions*

As in Chapter 2 we shall employ positive weight functions $Q_i(idea_i)$ that express the autonomous mental activity (e.g. introspection) of the idea at site i. Also positive acceptor functions $A_{i_1,i_2}(idea_1, idea_2)$ that express the strength of connections (associations, bindings) $i_1 \leftrightarrow i_2$ with the on- and off-values indicated by $idea_1$ and $idea_2$ respectively. The function pair $\{Q, A\}$ signifies the *personality* of the mind in question. With no essential restriction we can assume $A(i, i) = 1, \forall i$.

6.1.5   *Joint Probability Measure for Thoughts*

Our basic assumption is that the mind activity can be expressed by a *mind equation* that expresses the probability of a *thought*. Again, we borrow from GPT, p. 367.

$$p(thought) = 1/Z \prod_{i=1}^{n} Q_i(idea_i) \prod_{(i_1,i_2) \in \sigma} A_{i_1,i_2}(idea_{i_1}, idea_{i_2}) \qquad (6.6)$$

which is a variation of the Second Structure Formula in GPT. We shall denote the family of probability measures defined as in (6.6) by $\mathcal{SSF}$. We shall often deal with conditional probabilities that can be obtained by modifying (6.6). The formula (6.6) implies Markovian structure with respect to the graph σ. Compare this dependence with Figure 4 in Dean (2005). Note that the Q and A values depend upon the site number i and connection couple i_1, i_2 respectively. The Q-values indicate the sensitivity of the somas to input while the A-values measure the strength of synapses. This heterogeneity represents the personality characteristics of the mind being studied. This differs drastically from classical statistical mechanics where only one or a few types of units (atoms) are present and where, in contrast to the present study, the emphasis is on equilibrium situations. We should therefore not expect our system to behave like statistical mechanical systems.

The mind equation defining probabilities of thought can be given an interpretation in terms of the path integrals in Quantum Electro Dynamics; see Feynman, Kleinert (1986). Let σ be the graph describing the full network made up of sites s_i and connections c_{ij}. A thought can be represented by a subgraph $\sigma' \subset \sigma$, a path of sites and connections. Note that such a path differs from the standard notion in that it is not just a curve from an initial point to an end point, but can consist of several branches; we have seen several such paths in the earlier chapters. We shall introduce a *thought measure* μ with values given by the energies $\mu(s_i) = q_i$ and $\mu(c_{ij}) = a_{ij} + r$ with $r = log(\rho)$ as defined in Appendix B. We can then express the energy $E(thought)$ as the *path integral*

$$E(thought) = \oint_{t \in \sigma'} \mu(t)dt$$

where the "differential" dt consists of sites and connections, somas and dendrites. It is suggestive that the thought energy is built up by iterated sums of elementary energies along the path. The thought with the smallest path integral under given boundary conditions (current thought environment) is the most likely one, just as the least action in classical and quantum mechanics is the most likely one. We do not claim that this is more than an elegant analogy, but it is certainly thought provoking and deserves more attention.

It is convenient to use log-probabilities instead with

$$q(idea) = log[Q(idea)]; a(idea_1, idea_2) = log[A(idea_1, idea_2)];$$
$$z = log[Z] \tag{6.7}$$

so that equation (6.7) can be written in additive form

$$log[p(thought)] = -z + \sum_{i=1}^{n} q_i(idea_i)$$

$$+ \sum_{(i_1, i_2) \in \sigma} a_{i_1, i_2}(idea_{i_1}, idea_{i_2}). \tag{6.8}$$

Positive values of q and a indicate excitation and negative inhibition, respectively.

6.1.6 *Properties of SSF*

Then the conditional probability density of the thought $thought_{cond} = \sigma(idea_1, idea_2, \ldots, idea_n)$ given $cond = \{idea_{i_1} = \gamma_1, idea_{i_2} = \gamma_2, \ldots\}$, fixing the γ's, can be written as

$$p(idea_1, idea_2, \ldots, idea_n | cond)$$

$$= 1/Z_{cond} \prod_{i=1}^{n} Q_i(idea_i) \prod_{(i_1, i_2) \in \sigma} A_{i_1, i_2}(idea_{i_1}, idea_{i_2}) \tag{6.9}$$

but where some of the ideas are fixed to γ's and Z_{cond} is a new normalizing constant.

In other words: *the set $\mathcal{SSF}$ is closed under conditioning.*

On the other hand, if we want to find the marginal probability density of the sub-thought obtained by deleting $idea_{j_1}, idea_{j_2}, \ldots$ from *thought*, we get the sum

$$1/Z \sum_{idea_{j_1}, idea_{j_2}, \ldots \in V} \prod_{i=1}^{n} Q_i(idea_i) \prod_{(i_1, i_2) \in \sigma} A_{i_1, i_2}(idea_{i_1}, idea_{i_2}). \quad (6.10)$$

It will be sufficient to illustrate this for $j_1 = 1, j_2 = 2$. Then the product in (6.10) will have factors over the (i_1, i_2)-pairs

$$(1, 2), (2, 1), (1, compl), (compl, 1), (2, compl),$$

$$(compl, 2), (compl, compl) \quad (6.11)$$

with $compl = (3, 4, \ldots, n)$. Now we should multiply over the V-values of $idea_1, idea_2$. Note that the result will not always be the product of functions depending upon two variables; marginalization can bring us outside $\mathcal{SSF}$.

CONCLUSION: *MIND achieves conceptual inference via conditional probabilities.*

6.1.7 *Updating the Personality*

As time goes on, the personality $\{Q, A\}$ is affected by the mental activity that has occurred in current and new thoughts. More precisely, we shall assume the updating scheme for the time interval $(t, t+1)$.

$$q(idea_i, value) \rightarrow q(idea_i, value)$$

$$+ \epsilon_1 \text{ if } idea_i \in thought \ v \text{ on}; q(g_i, value) - \epsilon_2 \text{ else} \quad (6.12)$$

with $\epsilon_1 > 0; \epsilon_2 > 0; \epsilon_1 \gg \epsilon_2; v = V$-value. The first case consolidates the occurrence of $idea_i$ on; the second does the opposite but at a slower rate.

Also

$$a(i_1, i_2, value_1, value_2) \rightarrow a(i_1, i_2, value_1, value_2)$$

$$+ \delta_1 \text{ if } \sigma \text{ connects } i_1 \leftrightarrow i_2 \quad (6.13)$$

with the values indicated. Otherwise subtract δ_2 with $\delta_1 > 0, \delta_2 > 0$; $\delta_1 \gg \delta_2$.

6.1.8 *Processing Thoughts*

To simulate equation (6.17) we shall use stochastic relaxation in the form
of Markov Chain Monte Carlo (MCMC). This should be compared to the
statement at the end of Section 2 in Mumford (1992). We then need conditional probabilities

$$p(g_m|g_1, g_2, \ldots, g_{m-1}, g_{m+1}, \ldots, g_n) = \frac{p(g_1, g_2, \ldots, g_n)}{p(g_1, g_2, \ldots, g_{m-1}, g_{m+1}, \ldots, g_n)}.$$
(6.14)

Using equation (6.14) and canceling out lots of factors in the numerator
and denominator, including the partition function Z, this probability can be
written as N/D with the denominator independent of g_m. The numerator is

$$N = \prod_{i=1}^{n} Q_i(g_i) \prod_{(i,i') \in \sigma} A_{i,i'}[g_i, g_{i'}].$$
(6.15)

This gives us

$$N/D = \frac{1}{Z_m} Q_i(g_m) \prod_{(i,i') \in \sigma^-} A_{i,i'}[g_i, g_{i'}]$$
(6.16)

where σ^- means the graph consisting of the site m together with the bonds
emanating from it, and Z_m is a new normalizing constant.

Note that (6.16) can be rewritten as

$$log[p(g_m|g_1, g_2, \ldots, g_{m-1}, g_{m+1}, \ldots, g_n)]$$
$$= -z_m + q_i(g_m) + \sum_{(i,i') \in \sigma^-} a[g_i, g_{i'}]$$
(6.17)

with $z_m = log[Z_m]$. Using (3.2) with a straightforward threshold logic we
get a deterministic processing scheme.

We now introduce an alternative and more general definition: *An idea
is a family of probability measures over a set of binary on/off values on a
subset of nodes of the network.* This is for future reference only.

Equation (6.17) makes it clear that the mind equation is directly
related to the McCulloch-Pitts celebrated model for neurons with additive
inputs.

6.2 A Network Example

Let us look at an example. It is ridiculously small but will serve to illustrate what has been said above.

Starting from the set of ideas in Figure 6.4 we have chosen Q and A somewhat arbitrarily and applied a threshold logic to get a deterministic version of the thought process. If the sensors send signals that turn $idea_1, idea_2, \ldots$ on, this affects the joint probability density in (3.2) leading to a conditional probability density $p(thought|idea_1, idea_2, \ldots)$ easily obtained by setting these values to *on* and modifying the normalizing constant Z. Let us do this by starting MCMC for this density and the input [*bark, coarse*]; see Figure 6.4, top panel.

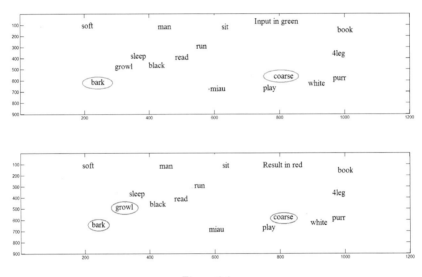

Figure 6.4:

The MCMC algorithm leads to the mind state with the ideas [*bark, coarse, growl*] shown in the bottom panel, so that the mind has tried to reconcile the input with the personality profile [Q, A] and found a likely thought appearing in the form of the template *dog*. It has extrapolated the input to the full concept *dog*.

On the other hand, if the sensory input is only *coarse* the result, see Figure 6.5, is just the same as the input. Obviously the evidence presented

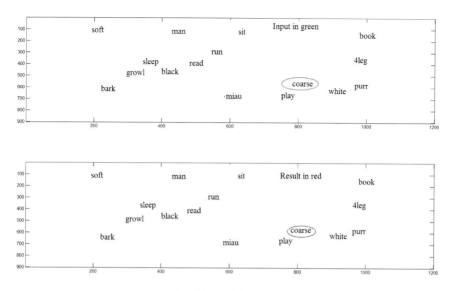

Figure 6.5:

by the senses was insufficient to allow an inference to any concept represented by the full thought template *dog*.

Behind this there may be lurking a threshold theorem of the type that von Neumann (1956) proved: if the deformed thought template is densely connected it is highly likely that a limited input information is enough to light up the whole thought template.

But a fuller input, *coarse, bark, purr, miau* confuses the mind: the result consists of the two concepts *cat* and *dog*, represented by one idea template each, together with *sit, run, black, white,* 4*leg* in Figure 6.6. The latter five ideas had only weak connections to the *dog, cat* concepts but enough for this (incorrect!) inference. Human thought is fallible.

Once the personality profile has developed and taken form (temporarily), how does A determine the template ideas? Let us make the MIND a pseudo-metric space by defining "distance" by

$$dist(idea_i, idea_i) = 0;$$

$$dist(idea_i, idea_j) = \frac{A_{ij}(0,1) + A_{ij}(1,0)}{A_{ij}(0,0) + A_{ij}(0,1) + A_{ij}(1,0) + A_{ij}(1,1)}, i \neq j.$$

$$(6.18)$$

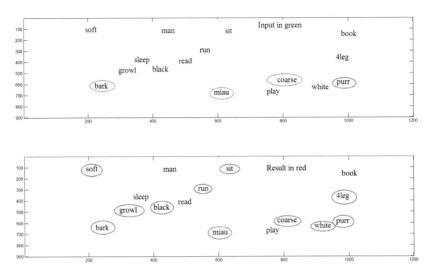

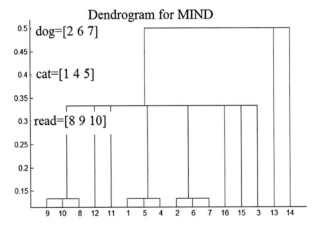

Figure 6.6:

Figure 6.7:

With this definition we can search for the ideas that cluster together, and are strongly linked. Using MATLAB's function "linkage" we get the hierarchical clustering in Figure 6.7.

Note that the clustering corresponds to the templates "dog", "cat", "read" in Figure 6.3 so that the idea templates have been "discovered" by the algorithm.

In this approach *encapsulation* was one of the main mental operations: combining the ideas in a thought to form a new unit, an elementary idea. Iterating this procedure this enables the thinker to operate on a more abstract plane. It characterizes abstract thinking on increasing levels of encapsulation. This was about the processes in the mind. But how about the neural correlate of the encapsulation activity?

6.2.1 *Concept Formation on the Neural Level*

Let us remind ourselves of the second structure formula of general pattern theory. It takes the form of the *mind equation*:

$$p(thought) = \frac{\kappa_n}{n! Z(T)} \prod_{i=1}^{n} Q(g_i) \prod_{(k,k') \in \sigma} A^{1/T}[b_j(g_i), b_{j'}(g_{i'})].$$

$$(6.19)$$

In this interpretation of equation (6.19) the A-values express the strength of the net connections, the Q-values the self activity of the nodes in the network. Hence $A \geq 1$ signifies an excitatory connection, while $A \leq 1$ means an inhibitory connection.

The term *p-clique*, probability clique, means a set of nodes most of which are connected with $A \gg 1$. This does not rule out that a few of the inner connections of the set have $A \ll 1$. Two *p*-cliques may overlap.

ASSUMPTION: *p-Cliques will be our network representation of concepts in a mind.*

6.3 Examples

To avoid foggy philosophizing we shall now argue by examples. Consider three concepts, apple, lemon, banana and represent them by configuration diagrams as we did earlier. Let us represent them by qualia, say, as the following attributes, chosen somewhat haphazardly,

apple = (surface, red, yellow, round, core, sweet, white, inner, seeds, black),

lemon = (surface, yellow, sour, rough, core, round, yellow, inner, seeds, white),

banana = (surface, yellow, oblong, smooth, white, seeds, black).

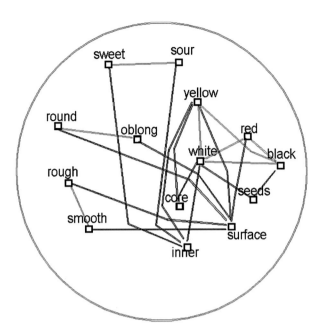

Figure 6.8:

In Figure 6.8 the sites (elementary ideas) are shown as small black squares and strong connection as black lines (excitatory) and red lines (inhibitory). There may also be weak connections but they will not be shown. This should be seen as only a small subset of a large neural net.

For the configuration diagrams in Figure 6.9 a green square means an active elementary idea, a green line an excitatory active connection, and red ones inhibitory active connections. Recalling the discussion earlier, the interpretation is that the green squares mean sensory inputs which turns on respective sites. In other words, *observations condition the probability measure*.

In Figure 6.12 we show how the concepts "apple", "banana" and "lemon" are formed into a higher abstraction "fruit".

6.3.1 The p-Cliques under Free Associations

This was for the case when inputs effect, or constrain, the thinking in terms of concepts. But what happens in the case of free association as discussed earlier? Let us consider a p-clique on n sites with the same

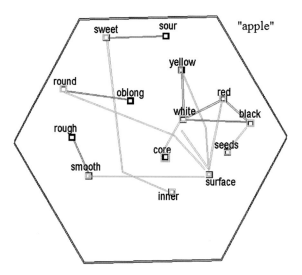

Figure 6.9:

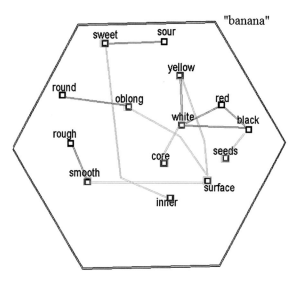

Figure 6.10:

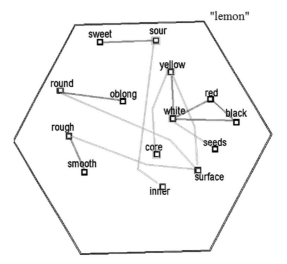

Figure 6.11:

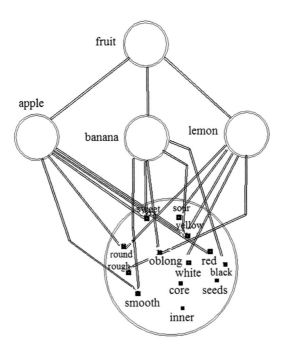

Figure 6.12:

136

A-values for on/off and, for simplicity, all Q-values $= 1$. Let $A1 = A(\text{on}, \text{on})$, $A12 = A(\text{on/off}) = A(\text{off}, \text{on})$, $A2 = A(\text{off}, \text{off})$. If the p-clique has x on-sites it has $1/2x(x-1)$ connections on/on, $x(n-x)$ on/off, and $1/2(n-x)(n-x-1)$ off/off sites. Consider $p(x) = Prob(\text{fields with } x \text{ on-sites})$. Using equation (3.5) and noting that some factors are constant we get the proportionality

$$p(x) \propto B(n,x)A1^{1/2x(x-1)} \times A12^{x(n-x)} \times A2^{(1/2(n-x)(n-x-1))}$$

$$(6.20)$$

where $B(n,x)$ is the binomial coefficient.

Keeping $A12$ and $A2$ constant we vary $A1$ starting with a small value. The size of the set of sites will then have a probability distribution as in Figure 6.13, so that the p-clique results in a full on-set fairly often, as well as nil occasionally, less often mixed. When we increase $A1$, the p-clique will more and more often result in a full on-set as seen in Figures 6.14–6.16.

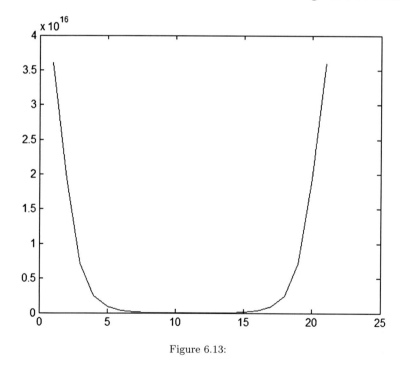

Figure 6.13:

This means that *the p-clique concisely represents a concept if the exhibitory connections are strong.*

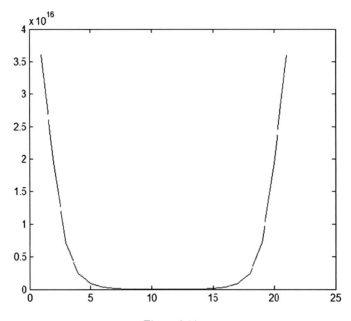

Figure 6.14:

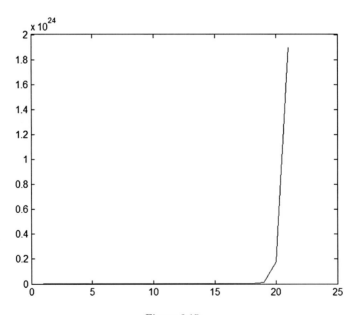

Figure 6.15:

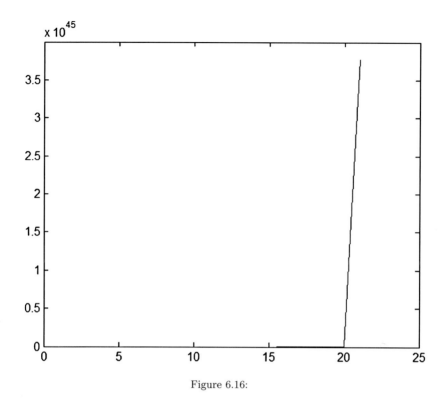

Figure 6.16:

6.3.2 *Kantian Thinking*

We have repeatedly referred to Immanuel Kant when he argued that human thought is essentially architectonic: starting with simple sensory inputs the thinker combines them into abstractions, then combines these into higher level abstractions, and so on. In this paper we have seen how these abstractions, or concepts, can be realized by neural-like nets. In an enormous network there will then be small subsets, p-cliques, representing elementary concepts. These subsets themselves cluster and new abstractions are formed, and so on, until the mind is able to perform high level thinking.

One of the *roles of concept formation is to increase efficiency* when the connections have been strongly enforced. For example, it is clear that learning the multiplication table, and storing it as an abstraction, must be more efficient than to carry out each multiplication anew from scratch. This efficiency lies behind the power of abstract thinking.

This has been just a thought experiment, but it may be possible to find empirical support. Indeed, one should try to register thoughts, preferably simple thoughts in a child, and analyze them. Of course, this is made difficult since linguistic utterances may conform poorly with the subjects thought processes, so that sophisticated experiments are required.

CONCLUSION: *The MIND can be implemented by biocomputing.*

6.4   Multi-Level Architecture

Let us look more closely at situations requiring several inter-connected levels. Say that we ask that the net realize the *thought* = "John say apple taste good". The following sequence of slides will do the job, but many alternative solutions are possible.

The "neural" connections are generated by simulating a Wiener bridge, i.e. a Wiener process bound down at the endpoints. This explains the irregular shape of the connections.

First look at the intra-connections at level $l = 1$ using a probability density of the form discussed above in Figure 6.17 and connections on levels $1, 2$ in Figure 6.18 and on $l = 1, 2, 3$, Figure 6.19.

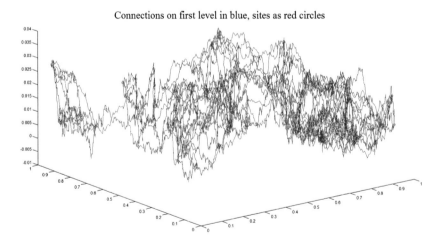

Figure 6.17:

Connections on second level in blue, sites as red circles

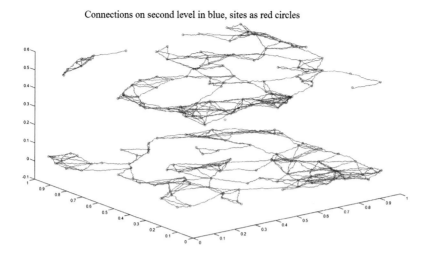

Figure 6.18:

Connections on third level in blue, sites as red circles

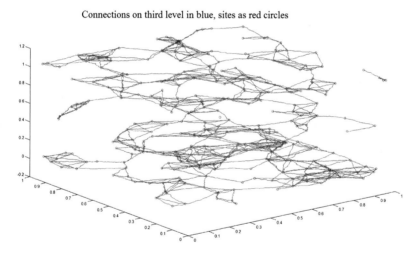

Figure 6.19:

Now look at the five templates

$$T1 = \text{“apple”}, \ T2 = \text{“good”}, \ T3 = \text{“taste”}, \ T4 = \text{“John”}, \ T5 = \text{“say”}$$
$$\tag{6.21}$$

as shown in Figures 6.20–6.24.

Template 1 in red = "apple"

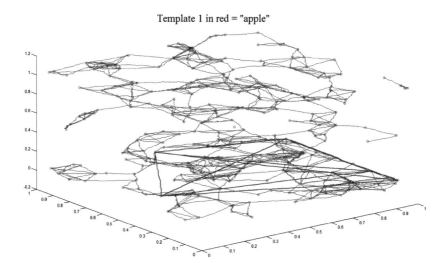

Figure 6.20: "apple"

Template T2 in cyan = "good"

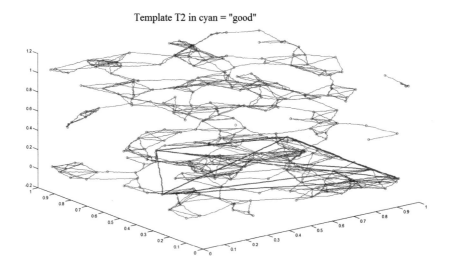

Figure 6.21: "good"

Template T3 in green = "taste"

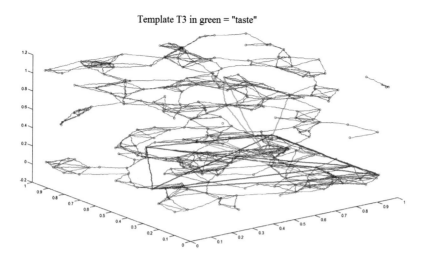

Figure 6.22: "taste"

T4 = "John" in black

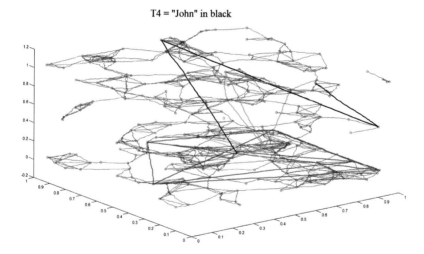

Figure 6.23: "John"

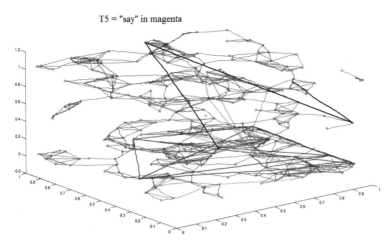

Figure 6.24: "say"

Finally, the inter-connections between the levels $l = 1, 2, 3$, and so on (see Figures 6.25 and 6.26).

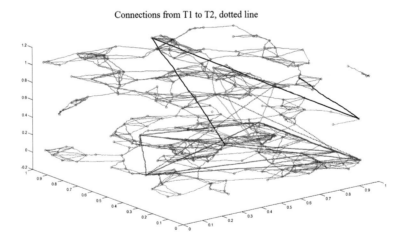

Figure 6.25: $T1$ to T

This example is of course extremely limited but it shows clearly how involved the network structure has to be in order that it be able to perform a simple mental task. Real neural networks have to be much more complex, say with 10^{10} units. What sort of mathematical tools do we have to deal

Connections from T1 to T3; dotted line

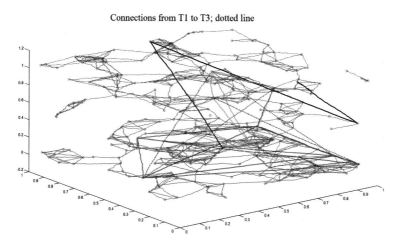

Figure 6.26: $T1$ to $T3$

with such overwhelming complexity? The methods of statistical mechanics are clearly insufficient as we have argued. But what else is there?

Before leaving this topic let us look at a related example but from another point of view, "things" to learn such as "car", ... Let us apply the updating that was introduced earlier but specialized to the rule

$$A(t+1, i_1, i_2) = min(A(t, i_1, i_2) * 1.1, 10) \qquad (6.22)$$

if sites (i_1, i_2) are connected at time t, otherwise

$$A(t+1, i_1, i_2) = A(t, i_1, i_2) * .99. \qquad (6.23)$$

We show some sequential "frames" of MIND and so on (see Figures 6.27–6.29).

After a run of 40 frames the acceptor matrix A corresponds to a Boolean incidence matrix $B = A > 8$ displayed in Figure 6.30.

Note that during the run all the information collected is in the updated acceptor matrix A. This means that the information is in the form of pairwise couplings. Nevertheless the MIND can learn the new concepts by searching for the cliques[1] associated with A.

---

[1] A clique is a subset of a graph such that all nodes in the set are connected to all other nodes in the set. A maximal clique is a clique that is not a proper subset of any other clique.

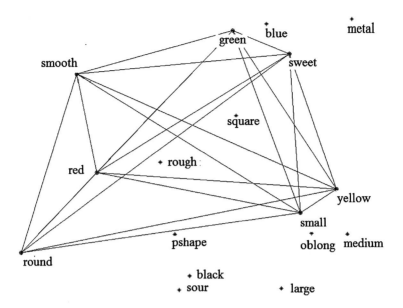

Figure 6.27: Sensor clique for "apple"

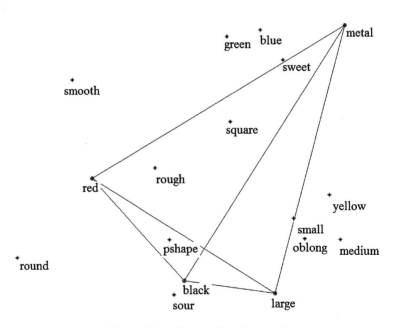

Figure 6.28: Sensor clique for "car"

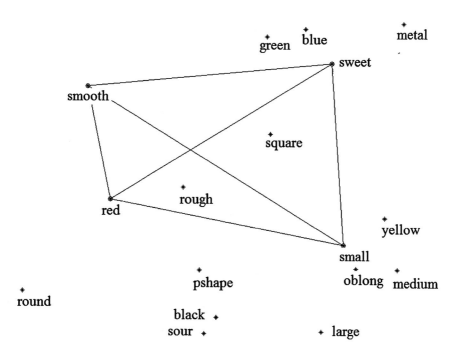

Figure 6.29: Sensor clique for "cherry"

Thus, to find the maximal cliques of this graph we use the MATLAB program "maximalCliques" and get a number of maximal cliques. It depends upon how many iterations we run. With short runs we find only some of the cliques, with many runs we find all. One result is the following set of observed cliques, see Figures 6.31–6.33.

All the accepted cliques are there, but in addition, in an observed frame there is one more. How should such an occurrence be explained? We leave that to the reader. Anyway, we have arrived at the

CONCLUSION: *Rational thinking consists of the manipulations of maximal cliques,*

and

CONCLUSION: *Concepts in the environment of the MIND correspond to maximal cliques in the associated network.*

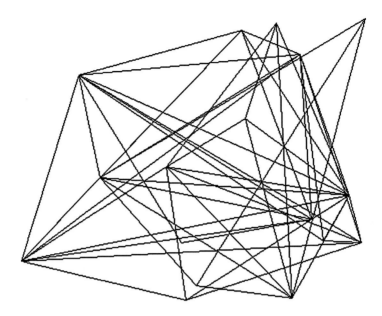

Figure 6.30: *A*-graph

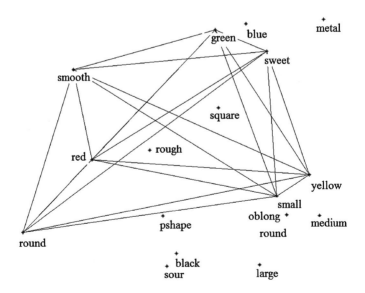

Figure 6.31: Observed clique

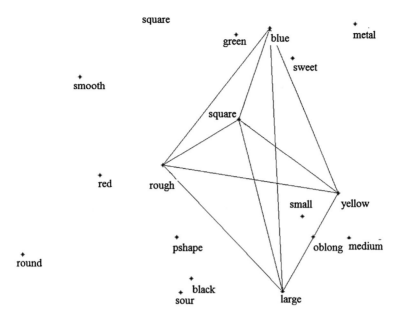

Figure 6.32:   Observed clique

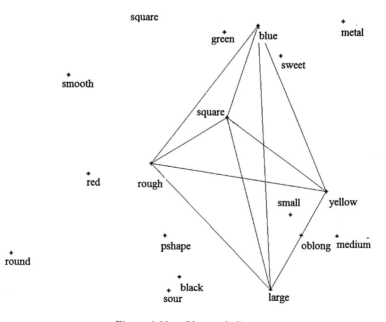

Figure 6.33:   Observed clique

6.4.1 *Manipulating Maximal Cliques*

Due to the importance of maximal cliques for human thought it is advisable to look more closely at their behavior. They can certainly overlap, see Figure 6.34, and give rise to two distinct but related ideas. The intersection *dog, bark, medium, rough,* red connectors, connects to *white,* with blue connectors, and to *black,* with green connectors. The intersection means "dog" without specifying color while the two maximal cliques indicate a "white" and "black" dog respectively.

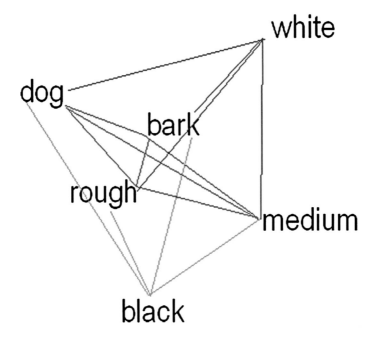

Figure 6.34:

It should be mentioned that the strength of the maximal cliques depends upon the values of Q and A. Concepts are not either-or but graduated in probability.

But given two minds, MIND1 and MIND2, how could they communicate growing up in more or less the same environment? Each of them have to develop some code of communication, perhaps in the form of gestures or, in more advanced cultures, as words. Say that MIND1 is exposed to apples with the sensory outputs $apple_1 = \{yellow, medium, round, sweet, sour\}$, let us code it x_1 in some code alphabet, but MIND2 only experiences

$apple_2 = \{yellow, medium, round, sour\}$. The diagram for $apple_1$ has the form of an inner clique to which the obtrusions *sweet* or *sour* are attached. The two minds communicating with each other will soon realize that x_1 does not correspond exactly to the code x_2; codes are not mapped bijectively into each other. To achieve bijectivity, and hence language understanding, MIND1 will have to introduce a new code x_1'. Continuing in this way, with the help of imitation, *language will evolve* based on success in communication as the optimality criterion.

We can expect to find mindscapes with many local minima close together.

Chapter 7

Reflections

In this chapter we shall allow ourselves to be less systematic, more free-wheeling, perhaps inviting protests from the reader. Let us take a step back and reflect on what we have done and not done. How is it related to earlier attempts to model the human mind in a formalized way? What is missing in the approach we have advocated and what should be pursued further?

7.1 Generalities on PoT

It is time to sit back and contemplate what we have achieved and what we have failed to do as we promised in Section 2.1.

7.1.1 *Raymondus Lullus and His Thinking Machine*

Nothing is new under the sun. Attempts to formalize human thinking can be traced back at least to Aristotle, see Appendix A.2, but there are many more in the far past. One of the most remarkable endeavors was that of Raymondus Lullus, Doctor Illuminatus, who in 1275 presented his "*Ars Combinatoria*", a machine that formalized thinking and was intended to prove or disprove fundamental, above all theological, statements. It consisted in one version of three circular disks, placed concentrically and could be turned around a common axis. One of them is shown in Figure 7.1.

The letters and other symbols on the disks should be interpreted as B = *Bonitas*, C = *Magnitudo*, D = *Duratio*, ... all attributes of God. By rotating the disks one gets combinations of ideas and thus theological statements. In this way Lullus was convinced that he could prove the existence of God

PRIMA FIGVRA.

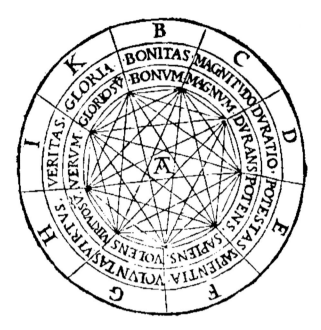

Figure 7.1:

and other fundamentals of the Christian faith. No doubt he was surprised that the Muslims in North Africa, where he traveled at the end of his life, did not accept these, to him, obvious truths; indeed they executed him as an infidel!

In spite of Lullus' failure to convince others, his attempt is an impressive attempt to formalize thought. It also led to computing machines via Leibniz, Pascal and Babbage, but that is another story. He *combined elements of thinking* in a way that is related, with many differences, to the approach of this book. The results of his derivations can be given in the form of graphs just as for *thoughts*.

We cannot leave this topic without mentioning the connection to the art of memory. *Ars Memorativa* was a respected discipline in the Middle Ages as a part of Rhetoric, and presented many mnemotechnical tools to facilitate remembering submerged memories. Partly influenced by Lullus it presented graphs like those in Figure 7.2. The interpretation of those graphs is similar to that of the circular diagrams employed by Lullus in Figure 7.1.

Figure 7.2:

Let us give an example how we would organize remembering in terms of our concepts, in particular using graphs labeled with ideas. Say that MIND is faced with the problem of finding a missing report, to find where it has been put.

To describe the cognitive environment of this MIND consider Figure 7.3 in which we see three elementary ideas "report", "placing" and

155

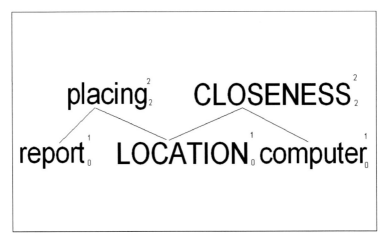

Figure 7.3:

"computer". Further the two modalities "LOCATION" and "CLOSENESS" with

MODALITY

$$\{i_1 = closeness_{garage,car}, i_2 = closeness_{comp,desk},$$
$$i_3 = closeness_{book,study}, \ldots\}, \qquad (7.1)$$

$$LOCATION = \{garage, \ desk, \ comp, \ldots, \ldots\}. \qquad (7.2)$$

All values of the acceptor matrix A shall be small in the above modalities except for $A(i_1, garage) \gg 1, A(i_1, car) \gg 1$ and $A(i_2, comp) \gg 1$, $A(i_2, desk) \gg 1$) and so on. If the MIND gets information that the report was close to the computer an argument with conditional probabilities is likely to infer that the report is on the desk. However, if $A(placing, desk)) \ll 1$ the inference is likely to change.

In this way remembering can be seen to be similar to the ancient methods of *Ars Memorativa* and its diagrams. The ingenious Lullus design was intended for theological thinking but nothing prevents us from replacing Bonitas, etc. by general elementary ideas. If we also allow dependencies less restricted than the circular ones in *Ars Combinatoria*, we arrive at a powerful thinking machine. This is just what we have done in the preceding chapter. Thinking consists, essentially, in combining ideas. Thus:

CONCLUSION: *Intelligence is the ability to connect ideas.*

7.1.2 Thinking vs. Language

Many readers will have thought that this approach is the same as for grammar in language. After all the thought diagrams are reminiscent of the parsing of sentences. Not so. Indeed, as we have pointed out repeatedly, thinking comes before language. Primates can probably think about objects of interest in their world, but have little or no language ability. When de Saussure (1916) talked about the *arbitrariness of the sign,* he just expressed the fact that naming of objects (and activities) must be preceded by the consciousness of them: words are created to represent thoughts. See also Pinker (2007).

Further, the devices of the grammar of natural languages such as declination, conjugation, word order, ... Instead we would argue that the graph structure of grammatical parsing, say that of TREE, is a consequence of the graph structure of thinking, say that of POSET, partially ordered sets. It is not necessarily a conscious decision, but one based on implicit understanding of the way we think. Two graphs in Figure 7.4 illustrate the differences.

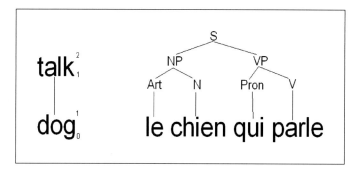

Figure 7.4:

There is some resemblance between them but they differ essentially both in topology and interpretation. A natural question is, however, how language has arisen to express thinking. We are not so presumptuous as to offer a solution to this mighty question, but let us reflect for a moment on possibilities. There is a small literature about finding the grammar of a language when a sample of sentences is presented. One elegant treatment of this can be found in Shrier (1977). It is restricted by the assumption that learning takes place in the presence of a teacher, supervised learning. This is acceptable since language is a social phenomenon. This should be taken with a grain of salt: in our view language originates in the last analysis from thinking. Therefore one could argue that the structure of many languages have a common genetic basis expressing the laws of thinking. The resulting linguistic similarities agree with the Chomsky doctrine.

The problem of constructing a communication code can have many solutions, but it is influenced by the response from the environment in which a MIND lives. Also, the topological structure of the thought process is mirrored in the form of the code. We shall give an example of a code that has strong similarities to some natural languages. Let us code a $thought = \sigma(idea_1, idea_2, \ldots, idea_n)$ by using a coding alphabet consisting of non-negative integers together with the separating symbols period "." and comma "," and "|". Code *thought* into the sequence $w_1|w_2|\ldots|w_i|\ldots|w_n$ of "words" w_i. For each integer i between 1 and n define the code word w_i as:

$$w_i = idea_i.in_{i1}.in_{i2}\ldots in_{ir_i}, out_{i1}.out_{i2}\ldots out_{is_i}. \tag{7.3}$$

Note the occurrence of the separating symbols "." and ",". Here $idea_i$ is the ith elementary idea in the *thought*, $in_{i1}.in_{i2}\ldots in_{ir_i}$ is the sequence of r_i i-values connecting down to $idea_i$, and $out_{i1}.out_{i2}\ldots out_{is_i}$ is the sequence of s_i i-values connecting up to $idea_i$.

This may sound complicated but is really quite natural as the following example illustrates. Say that we have a thought of size 5 as shown in the following table. Where the 5 rows enumerate the elementary ideas in *thought* in the first column. The second column enumerates the up-bonds and the third column the down-bonds.

John	4	0
book	4	0
Mary	4	0
give	5	1, 2, 3
yesterday	0	4

The entire code for *thought* will then be

$$code(thought) = w_1|w_2|w_3|w_4|w_5$$
$$= 314.4, 0|2007.4.0|168.4, 0|1226.5, 1\ 2\ 3|4881.0, 4$$

or in a more readable form,

$$code(thought) = w_1|w_2|w_3|w_4|w_5$$
$$= John.4, 0|book.4.0|Mary.4, 0|give.5, 1\ 2\ 3|yesterday.0, 4.$$

Compare with natural language where in some grammar w_1 denotes a proper noun "John" in nominative, w_2 a noun "book" in nominative and

the word w_3 the proper noun "Mary" in dative (in English using a preposi-
tional phrase). Further, w_4 is a transitive verb modified by the adverb "yes-
terday" in w_5. The devices declinations, conjugations, prepositions, word
order, intonation are all intended to express the connections in the graph
representing a thought.

The above code defines an *absolute language* in that it accommodates
all possible thought patterns expressed as above, meaning that it does not
need any additional rules or syntactic variables. This is in contrast to *relative
languages*, natural languages that adapt their grammars to the set of words
that particular groups of people use as labels. Note that it is denumerably
infinite, or, rather, potentially infinite. It may deserve further study —
perhaps as a tool in comparative linguistics.

In this connection one should mention literary analysis that has sometimes been based on studying linguistic phenomena. Instead it seems possible to observe the clouds of meaning in a text, clouds made up of unions
of modalities. This would change the emphasis from words to ideas, and in
some cases better express the substance of the thoughts behind the text.

7.1.3   *Questioning Introspection*

We have argued that introspection is a necessary tool for investigating the
human mind. Nevertheless, it has a distinct drawback. To illustrate let
us look at Figure 7.5. In *thought*(1) the MIND is hearing a dog bark.
In the next thought the MIND looks into itself, it is tired. The parallel (unconnected) earlier thought is still conscious. The result, *thought*(3),

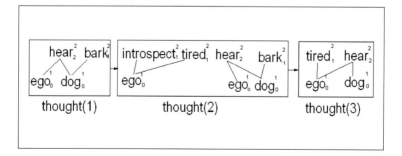

Figure 7.5:

is a deformed version of *thought*(1), the introspection has affected thinking. Observation distorts the mind activity. This will invalidate conclusions about the thought, but just as in quantum mechanics, this fact does not disqualify observation through introspection, at least on a macroscopic level.

7.1.4 *Thinking about the Unthinkable*

We have considered thoughts in MIND, both regular less frequently irregular, all the time generated by the available generator space G of available elementary ideas. But is there anything else? To shed a little light on this, we shall carry out a thought experiment.

Consider two minds, MIND1 = $\{G_1, \Sigma, Q_1, A_1\}$ and MIND2 = $\{G_2, \Sigma, Q_2, A_2\}$ with $G_1 \subset G_2$ so that MIND2 is mentally more powerful than MIND1. This implies that there is a set $MIND - DIFF = MIND22 \setminus MIND1 = MIND1 \cap MIND2^C$ of thoughts that can be created by MIND2 but not by MIND1. They are *unthinkable*. It is not only that MIND1 cannot think anything involving elementary ideas from $MIND - DIFF$; it is not even conscious of its mental limitation.

How can MIND1 overcome its limitation? There are two ways of doing it. Firstly, the *analytical knowledge acquisition* in which MIND1 applies the mind operations, the **mops** of Section 3.3, in particular *encapsulation*, to create new elementary ideas that can extend the thinking ability. Secondly, input from the external world may result in other new elementary ideas, *empirical knowledge acquisition*.

An adventurous reader will explore this exciting question further.

7.2 Substance in PoT

Referring back to the dichotomy substance-change in Section 1.2 we shall first reconsider Substance. An observant reader will have noticed that the choice of generator space G in GOLEM was quite arbitrary. In LEGACY it was done with more care but it is clear that more study is needed in how G should be selected in a systematic way. At the moment we can only offer a tentative suggestion with the hope that future researchers will pay more careful attention to this.

Let us consider the modality space $\mathcal{M}$ and its lattice of modalities. Organizing them into a tree structure and enumerating them going downwards and choosing left branches we get

$\mathcal{M} = \{Concrete, Abstract\}$
$Concrete = \{Animate, Inanimate\}$
$Animate = \{Fauna, Flora\}$
$Fauna = \{Human, Animal\}$
$Human = \{Body, BodyCovering, BodyInternal, Gender\}$
$Body = \{Health, Muscular, BodyParts\}$
$BodyCovering = \{Hair, Nails, Skin\}$
$BodyParts = \{Arms, Legs, Torso, Head\}$
$BodyInternal = \{Lungs, Gastro, Liver, Kidneys\}$
$HGender = \{HumanM, HumanF\}$
$HumanM = \{HumanMYoung, HumanMOld\}$
$HumanF = \{HumanFYoung, HumanFOld\}$
$Animal = \{Canine, Feline, Equinine, Insect, Bacterium, Virus\}$
$Flora = \{Tree, Flower, Fungus\}$
$Inanimate = \{Building, Food, Vehicle, Instrument, ArtObject,$
$\qquad\qquad\qquad Furniture, ReadingMaterial, Electronics\}$
$Abstract = \{Activities, Properties, Modes, Concepts\}$
$Activities = \{Work, Play, Relax, Move, Conflict, Amity\}$
$Work = \{ManualWork, IntellectualWork\}$
$Play = \{Sport, Game, PlayToy\}$
$Move = \{Walk, Drive, Bike, Swim\}$
$Conflict = \{Fight, Quarrel\}$
$Properties = \{Sensual, Asensual\}$
$Sensual = \{Visual, Auditory, Olfactory, Touch, Taste\}$
$Visual = \{Color, Size, Location, Orientation\}$
$Auditory = \{SoundHuman, SoundAnimal, SoundMechanical,$
$\qquad\qquad\qquad\qquad SoundMusic\}$
$SoundHuman = \{Singing, Talking, Crying, Snoring\}$
$SoundAnimal = \{Barking, Miawing, Neighing\}$
$SoundMusic = \{Classical, Jazz, Pop\}$
$SoundHuman = \{Singing, Talking, Crying, Snoring\}$
$Olfactory = \{SmellGood, SmellBad\}$
$Touch = \{TouchSoft, TouchHard\}$
$Taste = \{Sweet, Salty, Bitter\}$
$Asensual = \{Health, Happiness, Dreaming\}$
$Modes = \{When, Where, Why, How\}$
$Concepts = \{Love, Hate, Fear, Aggression\}.$

This list is embarrassingly incomplete and cannot serve as a blueprint for further work on software for LEGACY. It gives an idea, however, on how to construct a set of modalities. The modalities should then be filled with elementary ideas, for example

$$HumanMOld = \{Harry, John, \ldots, stranger, \ldots, patient, \ldots\}$$

of all old men in the environment of MIND.

The modalities are *universal entities*, common to most human minds in a certain cultural environment, while the elementary ideas may change from one individual to another. This is similar to Piaget's distinction between *general information* and *idiosyncratic information*. Therefore the modalities can be preset but the elementary ideas must be chosen separately for each individual.

How to do this efficiently is not clear. So far we have done this "manually", one after each other. This is a laborious process and one is likely to miss some important ideas. To automate this procedure the program would interrogate the user about the elementary ideas that should be introduced into the preprogrammed modalities. The response need not include arity, level and transfer information since this is already in the definition of the modalities. Nevertheless this seems cumbersome and could perhaps be facilitated by software devices; this remains to be done.

7.3 Change in PoT

7.3.1 *Continuity of Thinking*

The trains of thought form a stochastic process, a highly complicated one, but one that can be understood. Executing various versions of GOLEM it was noticed that the thoughts changed abruptly in time and little continuity was observed. Why was this?

An explanation is offered by a code fragment of the main function of GOLEM:

```
[content,connector]=delete_generator_connections_2(content,connector);
        see_mind(content,connector,number);pause(1);%new
[content,connector]=add_generator_up_Q(content,connector,theme);
        see_mind(content,connector,number);pause(1);%new

[content,connector]=add_generator_new(content,connector,Q_theme);
        see_mind(content,connector,number);pause(1);%new
```

```
[content,connector]=add_generator_up_Q(content,connector,theme);
      see_mind(content,connector,number);pause(1);%new
      close all
[content,connector]=add_generator_up_Q(content,connector,theme);
      see_mind(content,connector,number);pause(1);%new

[content,connector]=delete_generator_connections_2(content,connector)
      see_mind(content,connector,number);
      pause(1.6)

[content,connector]=delete_generator_connections_2(content,connector)
[content,connector]=add_generator_up_Q(content,connector,theme);
[content,connector]=delete_generator_connections_2(content,connector)
```

The "delete" and "add" statements occur frequently in this and other fragments of GOLEM, they obviously cause the trains of thought to exhibit little continuity. If this is deemed undesirable some of these could be commented out. This should lead to more continuity.

Another way of achieving the same goal is to apply the concept of distance between thoughts, *dist*, introduced in Section 3.2.1 and penalize the creation of $thought(t + 1)$ conditioned by $thought(t)$ for big values of $dist[thought(t + 1), thought(t)]$. Small consecutive values of this criterion will guarantee high continuity of the train of thoughts.

Chapter 8

Doubts and Certainties

Have the speculations in the previous chapters shed any light on how human thinking works? The author suffers no illusion about the way this work will be received by the cognoscenti in neural and cognitive science. They have the right to be skeptical — after all no empirical evidence has been suggested in favor of the thesis offered in the book. Doubt is good, it is the basic operating principle in science. Perhaps we should apply the Scottish verdict: Not proven.

However, introspection upon which the assumptions rest should not automatically be discarded in an extreme positivist attitude. It is observational with at least some limited possibility of replication by other researchers. And, as mentioned in Section 5.3 there may be future possibilities of comparing the performance of MIND with directly observed brain activities. However, we feel that we have proposed a cohesive theory with some credibility of how the human mind works. Therefore we dare suggest that

CONCLUSION: *The human mind can be understood without any metaphysical artifacts.*

****

14 BASIC RULES FOR THINKING ABOUT THINKING

1. *Thoughts are made up of discrete entities: ideas*
2. *Ideas are connected via bonds; defines semantic*
3. *The number of connections to an idea, the arity, is huge*
4. *Ideas constituting a thought are bound tightly together: a p-clique*
5. *Thought processes form a metric mind space*

$\boxed{6.}$ *Thinking is realized physically by a connected network*

$\boxed{7.}$ *The mind equation attributes strengths to ideas and connections between ideas*

$\boxed{8.}$ *The network structure implies that thinking is organized in terms of graphs*

$\boxed{9.}$ *Language, a small subset of thinking, must also be organized by graphs*

$\boxed{10.}$ *Thoughts are created probabilistically by the mind equation*

$\boxed{11.}$ *Thoughts are conditioned by boundary conditions*

$\boxed{12.}$ *Thoughts are concentrated to neighborhoods N(thought) in mind space*

$\boxed{13.}$ *The energy function E has a large number of local minima concentrating around thoughts*

$\boxed{14.}$ *The high level study of thinking should take place in mind space, not physical space*

****

References

There is an enormous literature on mind theories, especially general, informal ones, but also many mathematical/computational formalizations. Below we list only a small number of references that are directly related to the approach of this work.

J. Besag: Spatial interaction and the statistical analysis of lattice systems, J.R.S.S., 1974

G. Bell: A Personal Digital Score, Comm. ACM. 44, 2001

D. E. Brown: Human Universals, McGraw-Hill, 1991

N. Chomsky: Syntactic Structures, Mouton, The Hague, 1957

H. Cramer: Mathematical Methods of Statistics, Almqvist and Wiksell, 1946

A. R. Damasio: The Feeling of What Happens: Body and Emotion in the Making of Consciousness, Harcourt Brace and Comp., 1999

T. A. Dean: A Computational Model of the Cerebral Cortex. Proc. Twentieth National Conference on Artificial Intelligence, MIT Press, 2005

J.-L. Faulon: Automorphism partitioning, and canonical labeling can be solved in polynomial — time for molecular graphs, J. Chem. Inf. Comput. Sci., 1998

W. Feller: An Introduction to Probability Theory and its Applications, Volume I, 2nd Edition, Wiley, 1957

R. Feynman, H. Kleinert: Effective classical partition functions. Phys. Rev. A34, 1986

G. W. Gardiner: Handbook of Stochastic Models, Springer, 1990

J. Gottschall: Patterns of Characterization in Folk Tales Across Geographic Regions and Levels of Cultural Complexity: Literature as a Neglected Source of Quantitative Data. Human Nature 14 (365-382): 2003

U. Grenander: Lectures on Pattern Theory. Regular Structures Vol. III, (1981), Springer

U. Grenander: General Pattern Theory, Oxford University Press, 1993

U. Grenander: Windows on the World, CD-Rom, 2001

P. Hagmann, L. Cammoun, X. Gigandet, R. Meuli, C. J. Honey and van J. Wedeen, Olaf Sporns: Mapping the Structural Core of Human Cerebral Cortex, PLoS Biology, 2008

J. M. Hammersley and P. Clifford: Markov Fields on Finite Graphs and Lattices, preprint, University of California, Berkeley, 1968

O. R. Holsti: Content Analysis for the Social Sciences and Humanities, MA: Addison-Wesley, 1969

W. James: Varieties of Religious Experience, Dover Publications, 1902

I. Kant: Kritik der reinen Vernunft, Konigsberg, 1781

G. Mack: Interdisziplinare Systemtheorie, Lecture, University of Hamburg, 1998

E. Mally: Grundgesetze des Sollens, 1926

F. Mosteller and D. L. Wallace: Inference and Disputed Authorship, Center for the Study of Language and Information, 1964

D. Mumford: On the computational architecture of the neocortex, II: The role of cortico-cortical loops, Biological Cybernetics, 1992

W. S. McCulloc and W. Pitts: A logical calculus of the ideas immanent in nervous activity, Bull. of Math. Biophysics, 1943

J. von Neumann: Probabilistic logics and the synthesis of reliable organisms from unreliable components, "Automata Studies," edited by C. E. Shannon and J. McCarthy, Princeton University Press, 1956

B. Osborn: Parameter Estimation in Pattern Theory, Ph.D. thesis, Div. Appl. Math., Brown University, 1986

J. Pearl: Probabilistic Reasoning in Intelligent Systems, Morgan Kauffman, 1988

C. S. Peirce: On the algebra of logic; A contribution to the philosophy of notation, American Journal of Mathematics, 1885

S. Pinker: The Stuff of Thought, Viking, 2007

S. Pirmoradian: Software for LEGACY: http://people.sissa.it/pirmo rad/patterns_of_thought.html

V. Propp: Morphology of the Folktale. Trans., Laurence Scott, 2nd Edition, Austin: University of Texas Press, 1968

M. R. Quillian: Semantic memory. Minsky, M., Ed. Semantic Information Processing, pp. 216–270. Cambridge, Massachusetts, MIT Press, 1968

B. D. Ripley: Pattern Recognition and Neural Networks, Cambridge University Press, 1996

F. Rosenblatt: Principles of Neurodynamics: Perceptrons and the Theory of Brain Mechanisms, Spartan Books, 1962

F. de Saussure: Cours de linguistique generale, 1916

R. C. Schank: Conceptual Information Processing, North-Holland, 1975

S. Shrier: Abduction Algorithms for Grammar Discovery, Ph.D. thesis, Brown University, 1977

B. Spinoza: *Ethica ordine geometrico demonstrata*, 1670

O. Sporns and R. Kotter: Motifs in brain networks, PLoS Biology, 2, 2004

Y. Tarnopolsky: Molecules and Thoughts: Pattern Complexity and Evolution in Chemical Systems and the Mind, Rep. Pattern Theory Group at www.dam.brown.edu/ptg, 2003

Y. Tarnopolsky: spirospero.net/complexity.html

M. Tominaga, S. Miike, H. Uchida and T. Yokoi: Development of the EDR Concept Dictionary, Second Workshop on Japan-United Kingdom Bilateral Cooperative Research Programme on Computational Linguistics, UMIST, 1991

L. S. Vygotskij: Thought and Language, Cambridge, MA, MIT Press, 1962

J. B. Watson: Behavior: An Introduction to Comparative Psychology, 1914

J. Weizenbaum: ELIZA — A computer program for the study of natural language communication between man and machine, Communications of the ACM 9, 1966

L. Wittgenstein: Tractatus Logicus-Philosophicus, 6th Edition, London, 1955

B. L. Whorf: Language, Thought, and Reality. Selected Writings of Benjamin Lee Whorf, edited by J. B. Carroll, New York: MIT Press; London: John Wiley, 1956

R. Wille: Formal concept analysis, Electronic Notes in Discrete Mathematics, 2, 1999

WordNet, address http://wordnet.princeton.edu/

G. H. von Wright: An essay in deontic logic, MIND, 1968

Appendix A
Some Famous Mind Theories

Let us take a brief look at a few of the innumerable earlier attempts and see how they are related to the above discussion.

A.1 A Sample of Mind Theories

> L.R. Goldberg: *"We need to develop a structural model, some kind of an overarching taxonomy to link individual differences so that we're not all speaking idiosyncratic tongues."*

> BUT

> Paul Kline: *"The history of the psychology of personality, from Hippocrates onwards, is littered with the fragments of shattered typologies."*

Here is a list of some attempts to represent human thought. It is of course highly incomplete and the items are included only as pointers to what we have discussed in the previous sections. In spite of their different appearance they have elements in common with the research attitude presented in this work. The analogies may not be very strong. A more convincing parallel is to chemistry, something that Tarnopolsky has pointed out in a very convincing way; the reader may wish to consult Tarnopolsky (2003). The belief propagating systems in Pearl (1988) uses similar probabilistic concepts but with a different aim.

A.2   Syllogisms

Aristotle suggested syllogisms as guides for reasoning. Today it is difficult to see why they came to be considered so fundamental for thinking, but they were for a couple of thousand years, and innocent school children (including this author) were forced to memorize the possible syllogisms. Here is one of them

If all B's are A,
and all C's are B's,
then all C's are A.

Note the occurrence of the *variables* A, B, and C. They make the statement more general than would be a single instance of it, for example

all humans are mortal,

all Greeks are human,

then all Greeks are mortal

which is the special instance with A = "mortal", B = "human", C = "Greek". Compare with our use of modality abstraction.

A favorite syllogism among the Schoolmen, the so-called ontological proof that God exists:

If there was a God, He would be Perfect;

An aspect of Perfection is Existence;

Therefore, God Exists.

A.3   Formal Logics

Of greater interest is Boolean logic, introduced in Boole (1848), like $x \lor (y \land z)$, or in words "x or both y and z". Again, this is a generalization of $big \lor (little \land red)$. Another is predicate calculus, for example $\forall x(Ax \supset Bx)$, or in words "for all x it is true that if x is an A then x is a B". We want to mention that Peirce (1885), always original, actually used what is essentially graphs to represent some human thoughts; he called them existential graphs. Compare this to our use of configuration graphs!

Predicate calculus presumes Aristotelian syllogisms but is more powerful. Still more powerful logical systems of this type exist, but they have in common that they represent *exact thoughts*: the statements are true or false (at least this is the intention but caution is needed here) but less exact thinking is not represented by these systems. For example emotional thinking is not dealt with although this may actually be of greater human relevance for everyday use than exact reasoning. However, some philosophers have gone outside the classical domain of logical thought; as examples we mention Mally (1926) and von Wright (1968) and their studies of deontic logic.

A.4 Psychoanalysis

Emotional thinking is described by psychoanlysis as introduced by Siegmund Freud. Less formal than the above systems, this theory tries to understand the human mind in terms of elements: id, ego, superego, censor, libido, castration fear, child sexuality, transfer, repression, Oedipus complex, ... They are *combined* to form the nucleus of the mind of the patient, or at least the subconscious part of it, and are supposed to be discovered by the analyst through examination of dreams, slips, free associations and other expressions of the subconscious.

Among the many deviant practitioners of the psychoanalytic faith, Alfred Adler is one of the less exotic ones, actually representing more common sense than the other apostles. His "individual psychology" rejects Freud's original theories that mental disturbances were caused by sexual trauma, often in childhood, and he opposed the generalizations when dreams were interpreted, in most instances, as sexual wish fulfillment. Instead he used, as his basic elements of mind, feelings of inferiority, striving for power and domination, and wanted to understand mental activities as goal driven.

Posterity has not been kind to Freudian psychoanalytic theory, but it constitutes at least an audacious and admirable attempt to understand the human mind by representing them in terms of simple constituents. We also share this goal, but shall use more elemental units for building flexible models of thought. And we have of course strived for a quantitative theory employing probabilities.

A.5 Semantic Networks

The idea of semantic networks has been very popular in the AI community since its introduction in Quillian (1968). Such schemes are knowledge representation with nodes and directed connections between nodes. The nodes represent objects or concepts and the connections mean relations between nodes. A special case is the Petri net that has been suggested as a model of computation. Among other graph-based attempts we mention conceptual analysis, Wille (1999), and concept classification, Schanks (1975), Tominaga, Miike, Uchida, Yokoi (1991). A very ambitious attempt using objects and arrows can be found in Mack (1998).

We shall also use digraphs in our knowledge representations, but augmented in pattern theoretic terms, with not only generators and connectors, but also bond values, connection types, prior probability measures as well as algebraic operations on "thoughts". The semantic network was certainly a promising idea but interest in it seems to have waned in recent years. This may be due to the lack of specific structure in some of the work on semantic networks.

A.6 Formal Grammars

Following Chomsky (1957) many formal grammars have been suggested as models for human languages, for example context free grammars. They also use graphs, for example TREES, to generate the linguistic structures, but were intended to explicate language rather than thought. Among the systems mentioned here this one is closest in nature if not in details to the approach of this work, and this applies also to the current linguistic program Principles and Parameters. They differ above all in the distinction thought — language, in the author's opinion, a decisive opposition.

A.7 Associations

Behaviorism claims that human behavior can be explained in terms of stimulus-response associations, and that they are controlled by reinforcement. J. B. Watson (1914) described this approach in an influential book about human behavior. Mental terms like goal, desire, and will were excluded. Instead it used as building blocks the associations formed by repeated stimulated actions introducing couplings between input and output. Actually, we share the belief that association is a key element in human thought.

We shall also apply a compositional view, *but with many and very natural mental building blocks that represent extremely simple ideas.* They will be chosen as what seems to be natural and common sense entities in human thought, close to everyday life. Our choice of units is admittedly subjective but not wholly so. Indeed, we have been encouraged by the discussion of *human universals* in Brown (1991), who advocates the existence of universals organized into specific lists.

Appendix B

Consistency of Probability Measure

For the definition (3.2) to make sense as probabilities (normalized) we must have

$$Z(T) = \sum_{c \in \mathcal{C}(\mathcal{R})} \kappa_n \frac{1}{n!} \prod_{i=1}^{n} Q(g_i) \prod_{(k,k') \in \sigma} A^{1/T}[b_j(g_i), b_{j'}(g_{i'})] < \infty. \qquad (B.1)$$

This is similar to the condition for the probability measure over a stochastic CF language to be non-defective, see GPT 8.1.2. The above sum can be written as

$$\sum_{n=1}^{\infty} \kappa_n \sum_{c \in \mathcal{C}_n(\mathcal{R})} \frac{1}{n!} \prod_{i=1}^{n} Q(g_i) \prod_{(k,k') \in \sigma} A^{1/T}[b_j(g_i), b_{j'}(g_{i'})] \qquad (B.2)$$

where $\mathcal{C}\backslash(\mathcal{R})$ consists of all regular configurations of the mind of size n. If the maximum arity is ω_{max}, the cardinality of σ_n is bounded by

$$|\sigma| \leq (n\omega_{max})^n \qquad (B.3)$$

so that the above sum is bounded by

$$\sum_{n=1}^{\infty} \kappa_n \sum_{c \in \mathcal{C}_n(\mathcal{R})} \frac{1}{n!} \prod_{i=1}^{n} Q(g_i) \prod_{(k,k') \in \sigma} A^{1/T}[b_j(g_i), b_{j'}(g_{i'})]$$

$$\leq \sum_{n=1}^{\infty} \kappa_n (n\omega_{max})^n \frac{1}{n!} Q_{max}^n A_{max}^{n\omega_{max}}. \qquad (B.4)$$

In order that this series converge it is sufficient to ask that

$$\kappa_n = O(\rho^n); \rho < \frac{1}{e\omega_{max}Q_{max}A_{max}^{\omega_{max}}}. \qquad (B.5)$$

Indeed, this follows from the classical Stirling formula

$$n! \asymp \sqrt{2\pi n} \left(\frac{n}{e}\right)^n \tag{B.6}$$

which implies that the terms in the sum are dominated by those of a geometric series with ratio less than one if (B.5) is satisfied.

This means that we have the

PROPOSITION: *The probability measure is well defined if the combinatorial complexity of the mind is bounded by (B.5): the probability of large configurations representing complicated mental modes must be small enough.*

Otherwise the mind would expand indefinitely, taking on more and more complicated states, leading to a mental explosion.

We shall use the notation $\pi_n = \kappa_n/n!$ which describes the probabilities of the size of content (c). It should be noticed that (B.5) is satisfied with $\pi_n = Poisson_n(\mu)$, a Poisson distribution with mean $\rho = \mu$. It is not clear if this can be motivated by an underlying Poisson process in the MIND.

NOTE: In terms of Gibbsian thermodynamics the above is not the canonical ensemble. Indeed, the number of interacting elements is not fixed but random and variable. Thus we are dealing with Gibbs' *grand* canonical ensemble.

Appendix C
A Modality Lattice

Rectangles shall stand for modalities and diamond shapes for unions of modalities that do not form modalities themselves. Primitive ideas are shown under the rectangles.

The modularity lattice is too big to show in its entirety. Instead we show parts of it. The modality ANIMATE in Figure C.1 and BEHAVE, Figure C.2.

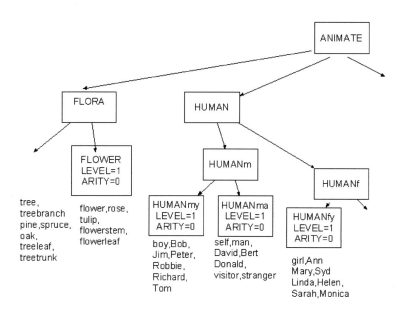

Figure C.1:

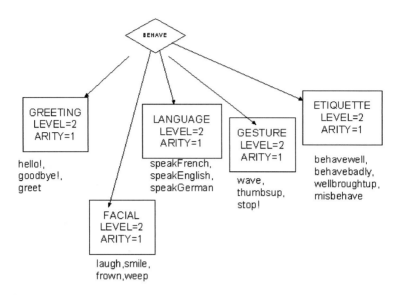

Figure C.2:

Note that **BEHAVE** is not a modality but is broken up into modalities. And INANIMATE, in Figure C.3, and the non-modality INVOLVEhum, Figure C.4.

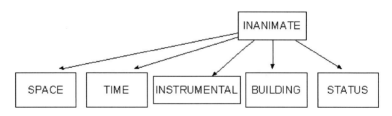

Figure C.3:

Finally **PERSON** is shown only in part, see Figure C.5.

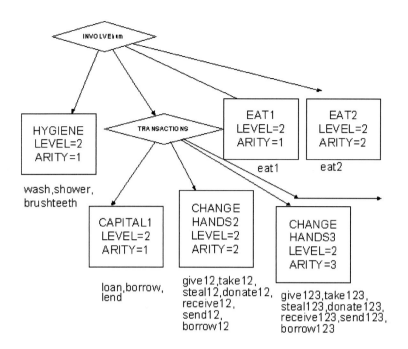

Figure C.4:

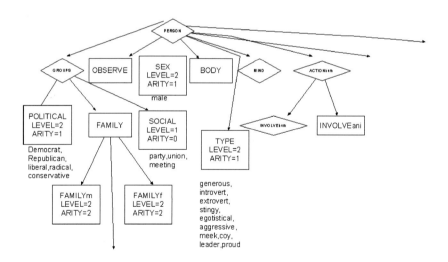

Figure C.5:

Appendix D

Dynamics of Random Process of Thinking

The fundamental probability measure used for describing the thought process had the density

$$p(thought) = \frac{\kappa_n}{n! Z(T)} \prod_{i=1}^{n} Q(idea_i) \prod_{i,j} A^{1/T}[(idea_i, idea_j)]. \quad \text{(D.1)}$$

To simplify notation we shall leave out the factor $n!$, absorbing it in κ_n. Also, let the intellectual temperature $=1$. It can be obtained as the limit of a dynamic scheme, see Section 3.4. To build thoughts in the Kantian sense from elementary ideas let us use building steps of four types in continuous time t:

(1) add a generator $idea_i$ with probability $\nu_1(idea_i)dt$ in a time interval $(t, t + dt)$,

(2) delete a generator $idea_i$ with probability $\nu_2(idea_i)dt$ in a time interval $(t, t + dt)$,

(3) add a connector $idea_i - idea_2$ with probability $\nu_3(idea_i, idea_j)dt$ in a time interval $(t, t + dt)$,

(4) delete a connector $idea_i - idea_2$ with probability $\nu_4(idea_i, idea_j)dt$ in a time interval $(t, t + dt)$.

Introduce a birth- and death-process with transition equation

$$p(thought(t + dt)) = P1 + P2 + P3 + P4, \quad \text{(D.2)}$$

$$P1 = \sum_{idea_i minus} p(thought')\nu_1(idea_i)dt \quad \text{(D.3)}$$

$$+ p(thought(t)) \left[1 - \sum_{idea_i minus} \nu_1(idea_i)dt \right] + o(dt),$$

$$P2 = \sum_{idea_i add} p(thought')\nu_2(idea_i)dt \tag{D.4}$$

$$+ p(thought(t)) \left[1 - \sum_{idea_i add} \nu_1(idea_i)dt \right] + o(dt),$$

$$P3 = \sum_{connector_{ij} minus} p(thought')\nu_3(idea_i, idea_j)dt \tag{D.5}$$

$$+ p(thought(t)) \left[1 - \sum_{connector_{ij} minus} \nu_3(idea_i, idea_j)dt \right] + o(dt),$$

$$P4 = \sum_{connector_{ij} add} p(thought')\nu_4(idea_i, idea_j)dt \tag{D.6}$$

$$+ p(thought(t)) \left[1 - \sum_{connector_{ij} add} \nu_4(idea_i, idea_j)dt \right] + o(dt).$$

To explain the notation look at equation (D.3). The first summation should be over those *thought'* equal to *thought* except that an idea $idea_i$ has been deleted. Similarly for the rest of the equations.

As $dt \downarrow 0$ we get the familiar differential equation

$$\frac{dp(thought, t)}{dt} = Q1 + Q2 + Q3 + Q4 \tag{D.7}$$

with

$$Q1 = \sum_{idea_i minus} p(thought')\nu_1(idea_i) \tag{D.8}$$

$$- p(thought(t)) \sum_{idea_i minus} \nu_1(idea_i),$$

$$Q2 = \sum_{idea_i add} p(thought')\nu_2(idea_i) \tag{D.9}$$

$$- p(thought(t)) \sum_{idea_i add} \nu_2(idea_i),$$

$$Q3 = \sum_{connector_{ij} minus} p(thought')\nu_3(idea_i, idea_j) \tag{D.10}$$

$$- p(thought(t)) \sum_{connector_{ij} minus} \nu_3(idea_i, idea_j),$$

$$Q4 = \sum_{connector_{ij}\,add} p(thought')\nu_4(idea_i, idea_j) \tag{D.11}$$

$$- p(thought(t)) \sum_{connector_{ij}\,add} \nu_3 4(idea_i, idea_j).$$

Now let us specify the birth and death intensities for the elementary ideas; compare with Section 3.2. Put

$$\nu_1(idea_i) = Q(idea_i)\frac{\kappa_{n+1}}{\kappa_n}, \tag{D.12}$$

$$\nu_2(idea_i) = 1/Q(idea_i)\frac{\kappa_{n-1}}{\kappa_n}, \tag{D.13}$$

$$\nu_3(idea_i, idea_j) = A(idea_i, idea_j), \tag{D.14}$$

$$\nu_4(idea_i, idea_j) = 1/A(idea_i, idea_j). \tag{D.15}$$

Direct calculations show that the density in equation (D.1) satisfies the equilibrium equation

$$\frac{dp(thought, t)}{dt} = 0. \tag{D.16}$$

Moreover we can verify that the MIND is in detailed balance.[1] Hence the MIND allows microscopic reversibility.

---

[1] See Gardiner (1990), pp. 148–165.

Appendix E

Software for GOLEM

Executing GOLEM calls a number of functions, first of all the main function "think".

E.1 Main Function

The output to "think" is of the form [content, connector]. The function loads a file "mind-data" containing a generator space, the modality lattice and much else; it should be placed in c:\mind_data. The code is complicated, but the reader is recommended to read it, at least briefly, in order to see what programming strategy has been applied. Otherwise it would be hard to figure out what devices have been used to build the code.

```
function think
%creates complete ''thought'' and displays 2-idea if there is one...
 in thought
%set seed forrandomness
rand('state',sum(100*clock));
c=menu('CHOOSE A MIND OPERATION','THINKING DRIVEN BY THEME',...
'CONTINUOUS THOUGHT','THINKING DRIVEN BY EXTERNAL INPUTS',...
    'FREE ASSOCIATIONS','SET PERSONALITY PROFILE',...
'SET MIND LINKAGES',...
'THE VISIBLE MIND','SEE CREATED IDEAS'.'DEVELOP');
switch c
```

The first case implements thinking in *themes*; it is one of the most important options:

```
    case 1
[content,connector]=think1;
hold on
load c:\mind_data
```

```
%is there a 2-idea?
cont=content(:,2);
mods=g_mod(cont)
gs= ismember(mods,180);
if any(gs)
    see_mind(content,connector)
    hold on
    blinktxt(.6,.7,'NOTE ABSTRACT IDEA')
    hold on
    pause(4)
    figure('Units','Normalized','Position',[0 0 1 1])
    axis off
    a=menu('ANALYZE IDEA ?','YES','NO')
    if a==1
        close all
        ind=find(gs);idea_generator=cont(ind(1));
idea_generator=G(idea_generator);
        idea_name=idea_generator.name;
        number=name_2_number(idea_name);
        idea_content=CREATION{1,number,1};
idea_connector=CREATION{1,number,2};
        see_mind(idea_content,idea_connector)
        N=radix2num(idea_content(:,2),r)
        text(.1,.7,['IDEA WITH GOEDEL NUMBER '...
,num2str(N)],'FontSize',30,'Color','b')
        pause
    end
    close all
    b=menu('APPLY ABSTRACTION OPERATOR TO IDEA ?','YES','NO')
    if b==1
        see_mind_mod(idea_content,idea_connector)
        pause
    end
end
c=clock;c=rem(c(5),5);
if c ==0
    [Q,A]=memory(content,connector);
    close all
    clf
figure('Units','Normalized','Position',[0 0 1 1])
axis off
text(.2, .2, ['STRENGTH OF MIND LINKAGES UPDATED'],...
'Fontsize',20','Color','b')
pause(1)
end
close all
```

The next case is more complicated. It deals with thinking where the trajectory jumps from one theme to another repeatedly and sometimes creates new ideas:

```
case 2
    load('C:\mind_data');
    %figure('Units','Normalized','Position',[0 0 1 1])
    %axis off
    clf
    answer=questdlg('MORE CONTINUOUS THOUGHT ?', 'YES','NO');
    if answer==2
        return
    end
    duration=menu(['HOW MANY SECONDS OF CONTINUOUS THOUGHT  ? '],...
'10','20','30','40');
    duration=duration*10;%duration=str2num(duration)
    t0=clock;genre_old=1;


    while etime(clock,t0)<duration
            genre=select(ones(1,9)./9)
            if ~(genre==genre_old)
                figure('Units','Normalized','Position',[0 0 1 1])
                axis off
                clf
                text(.01, .5,['MIND TRAJECTORY CHANGES DIRECTION'],...
'FontSize',26,'Color','y')
                axis off
                pause(.6)
            else
            end
            content=[];load c:\mind_data G
            %create thought germ ''content,connector''
            [content,connector]=think2(genre);
            [content,connector]=...
add_generator_up_Q(content,connector,genre);
        [content,connector]=...
add_generator_up_Q(content,connector,genre);
            %[content,connector,Q_theme]=build_thought_mod(genre);
            %see_mind_germ(content,[])
            pause(3)
            close all
            w=[];
            if isempty(content)
                figure('Units','Normalized','Position',[0 0 1 1])
                text(.2,.1 ,...
['EMPTY MIND'],'Color','r','FontSize',20)
                axis off
                pause(1)
            else
            v=content(:,2);n_v=length(v);
            k=1:n_v
```

```
              g=G(v(k));
              w=[w,g.level];
              if all(ismember(w,1))
                     figure('Units','Normalized','Position',[0 0 1 1])
                     text(.2,.1 ,['STOP THINKING!   NO OPEN BONDS!'],...
'Color','r','FontSize',20)
                     axis off
                     pause(1)
              end
              end
              %is any down bond open?
              found=1;
              while found==1
                     [i,h,omega,found]=...
find_open_down_bond(content,connector);
                     if found==0
                            see_mind(content,connector)
                            pause(1)
                            close all
                            %return here?
                     else
                            [content,connector,found]=...
connect_down_bond(content,connector, i,h,omega,Q_theme);
                            see_mind(content,connector)
                            pause(1)
                     end

              end
              see_mind(content,connector);
              pause(1.6)
              close
        [content,connector]=add_generator_up_Q(content...
,connector,genre);
        [content,connector]=add_generator_up_Q(content...
,connector,genre);
        see_mind(content,connector);
              pause(1.6)
              close

              [content,connector]=dom_thought(content,connector);
              see_mind_dom(content,connector);
              pause(3)
              genre_old=genre;
              close all
      end
  %now detect top_2ideas
    [top_2ideas_g,top_2ideas_h]=get_top_2ideas(content...
,connector); %these are the top_2ideas
    n_ideas=length(top_2ideas_g);
    ns=zeros(1,n_ideas);
```

```
if n_ideas ==0
    figure('Units','Normalized','Position',[0 0 1 1])
    axis off
    text(.2,.8,'No Conscious Thought','FontSize',32)
    text(.8,.1,['Press Enter to Continue'],'FontSize',8)
    return
end

for t=1:n_ideas
     gs=top_2ideas_g{1,t,:}; ns(t)=length(gs);
end
[Y,I]=max(ns);
m=I(1);
hs=top_2ideas_h{1,m,:};gs=top_2ideas_g{1,m,:};
content1(:,1)=hs';content1(:,2)=gs';n=length(hs);connector1=[];
for k1=1:n
    for k2=1:n
        for j=1:3
            h1=hs(k1);h2=hs(k2);g1=gs(k1);g2=gs(k2);
            segment=(connector(:,1)==h1)&(connector(:,2)==h2)...
&(connector(:,3)==j);
                if any(segment)&(g1~=g2)
                    connector1=[connector1;[h1,h2,j]];
                    else
                end
        end
    end
end
%add new idea to ''G''
r=length(G);n_new_ideas=length(gs_in_mod{180});...
%note numbering of ''new ideas'' modality
G(r+1).name=['<idea',num2str(n_new_ideas+1),'>'];
G(r+1).level=1;
G(r+1).modality=180;
g_mod=[g_mod,180];x=size(CREATION);
    n_new_idea=x(2);
    CREATION{1,n_new_idea+1,1}=content1;
    CREATION{1,n_new_idea+1,2}=connector1;
    Q=[Q,1];A_new=zeros(r+1);A_new(1:r,1:r)=A;

 A_new(r+1,:)=ones(1,r+1);A_new(:,r+1)=ones(r+1,1);A=A_new;
    figure('Units','Normalized','Position',[0 0 1 1])
    axis off
    text(.2,.8,'New Idea Created !','FontSize',32)
    text(.5,.1,['Press Enter to Continue'],'FontSize',20)
    %pause
    [L1,L2,L3,L4]=get_levels(G);
    clear content connector omega genre theme
    clear content1 connector1
    save c:\mind_data
```

The third case accepts inputs from the external world and learns from experience by updating "*Q*" and "*A*":

```
case 3
    %get input from external world:
%carries out inference from inputted thought
load c:\mind_data
external_world=sensory;
l_external=length(external_world);connector=[];content=[];
%now start to build internal MIND as configuration
content_col2=[];connector1=[];l=0;
for nu=1:l_external
    sub=external_world{nu};
    l_sub=length(sub(:,1));content1=zeros(l_sub,2);connector1=[];
    content1(:,1)=[l+1:l+l_sub]';content1(:,2)=sub(:,2);
    [content1,connector1]=add_connector_new(content1,connector1);
    connector=[connector;connector1];
    content_col2=[content_col2,sub(:,2)'];
    l=l+l_sub;
end
    l_scene=length(content_col2);
    content=zeros(l_scene,2);content(:,1)=[1:l_scene]';
content(:,2)=content_col2';
    see_mind(content,connector)
    pause(3)
    close
    v=content(:,2);n_v=length(v);w=[];
for k=1:n_v
g=G(v(k));
w=[w,g.level]
end
if all(ismember(w,1))
    figure('Units','Normalized','Position',[0 0 1 1])
    text(.2,.1 ,['STOP THINKING!   NO OPEN BONDS!'],...
'Color','r','FontSize',32)
    axis off
    pause(1)
    return
end

    figure('Units','Normalized','Position',[0 0 1 1])
    axis off
    text(0,.5,['Input complete. Press Enter...
 to continue and wait...'],'FontSize',22)
    pause
    close all
for iter=1:3
    [content,connector]=add_generator_up(content,connector);
    [content,connector]=add_generator_down(content,connector);
end
```

```
%is any down bond open?
found=1;
while found==1
    [i,h,omega,found]=find_open_down_bond(content,connector);
    if found==0
            see_mind(content,connector)
            pause(1)
            close all
            %return
    elseif found==1
        Q(gs)=20;Q_theme=Q;
        [content,connector,found]=connect_down_bond(content,...
connector, i,h,omega,Q_theme);
        see_mind(content,connector)
        pause(1)
    end
end
see_mind_infer(content,connector)
    close all
    [Q,A]=memory(content,connector);

close all
```

In case 4 the thinking is not controlled by either external inputs nor by thematic restrictions. The result is very chaotic thoughts:

```
case 4
    %free associations
figure('Units','Normalized','Position',[0 0 1 1])
    load('C:\mind_data');
    text(.2,.5,['WAIT...'],'FontSize',32)
    axis off
    pause(1);content=[],connector=[];
    n_input=0;
    sto=1;
    while sto==1
    for iter=1:3
        [content,connector]=add_generator_new(content,connector);
    end
    see_mind(content,connector)
    text(.1,.98,'CHAOTIC THINKING...','Fontsize',20,'Color','y')
pause(1)
close
for iter=1:4
    [content,connector]=add_generator_up(content,connector);
    [content,connector]=add_generator_down(content,connector);
see_mind(content,connector)
text(.1,.98,'CHAOTIC THINKING...','Fontsize',20,'Color','y')
pause(1)
end
```

```
for iter=1:1
[content,connector]=delete_generator_connections(content,connector);
end
pause(1)
close
[content,connector]=see_mind_dom(content,connector)
text(.1,.98,'CHAOTIC THINKING...','Fontsize',20,'Color','y')
hold on
text(.2,.05,'Press ENTER to continue', 'FontSize',12)
hold off
pause
close all
   figure('Units','Normalized','Position',[0 0 1 1])
   axis off
   q=menu('CONCENTRATED THOUGHT ? HARD THINKING,...
 TAKES TIME...WAIT...', 'YES','NO');
if q==1
  [content,connector]=add_connector_new(content,connector) %note;
  see_mind(content,connector)
  hold on
text(.2,.05,'Press ENTER to continue', 'FontSize',12)
pause
   close
end
   figure('Units','Normalized','Position',[0 0 1 1])
   axis off
   p=menu('CONTINUE WITH FREE ASSOCIATIONS ?', 'YES','NO');
   if p==2
       sto=2;
   see_mind(content,connector)
   hold on
text(.2,.05,'Press ENTER to continue', 'FontSize',12)
hold off
pause
close all
     end
end
[Q,A]=memory(content,connector);
figure('Units','Normalized','Position',[0 0 1 1])
axis off
text(.1, .5, ['MIND LINKAGES UPDATED: FORGET AND REMEMBER'],...
'Fontsize',20','Color','b')
pause(1)
close all
```

The next case lets the user define a personality profile for "self":

```
case 5
    set_personality
```

Case 6 implements the personality profile by changing "*Q*" and "*A*":

```
case 6
    load c:\new
    load c:\mind_data
    %personality_behavior are sets of g's
    %first set Q's
    r=length(G);
for g=1:r
    if strcmp(G(g).name,'self')
        sel=g;
    end
end
 A(greedy,sel)=(1-val1)*3;
    A(generous,sel)=val1*3;
     A(scholastic,sel)=(1-val2)*3;
    A(athletic,sel)=val2*3;
     A(aggressive,sel)=(1-val3)*3;
    A(mild,sel)=val3*3;
     A(selfish,sel)=(1-val4)*3;
    A(altruistic,sel)=val4*3;
    %symmetrize
    A=(A+A')./2;
    save c:\mind_data
    figure('Units','Normalized','Position',[0 0 1 1])
    axis off
    text(.1,.9,'STRENGTH OF MIND LINKAGES SET TO:...
 ','Color','y','Fontsize',28)
    text(.1,.8,['greedy: ',num2str(1-val1)],...
'Color','y','Fontsize',28)
    text(.1,.7,['generous: ',num2str(val1)],...
'Color','y','Fontsize',28)
    text(.1,.6,['scholastic: ',num2str(1-val2)],...
'Color','y','Fontsize',28)
    text(.1,.5,['athletic: ',num2str(val2)],....
'Color','y','Fontsize',28)
    text(.1,.4,['aggressive: ',num2str(1-val3)],...
'Color','y','Fontsize',28)
    text(.1,.3,['mild: ',num2str(val3)],...
'Color','y','Fontsize',28)
    text(.1,.2,['selfish: ',num2str(1-val4)],...
'Color','y','Fontsize',28)
    text(.1,.1,['altruistic: ',num2str(val4)],...
'Color','y','Fontsize',28)
    pause
```

In case 7 the MIND is displayed as connections between elementary ideas situated on the circumference of a circle. Note the idea "self" as a small red star:

```
case 7
    %display ''A'' linkages
```

```
        load c:\mind_data G A r
        angles=2*pi.*[0:r-1]./r;
        xs=cos(angles);ys=sin(angles);
        figure('Units','Normalized','Position',[0 0 1 1])
        axis off
        text(.3,.8, 'VISIBLE MIND','Fontsize', 25,'Color','r')
        text(.3,.6, 'LOCATION OF ''SELF'' INDICATED BY *'...
,'Fontsize', 25,'Color','r')
        text(.3,.4, 'WAIT !','Fontsize', 25,'Color','r')
        text(.3,.2, 'TAKES A WHILE...','Fontsize', 25,'Color','r')
        pause(2)
        close all
        clf
        figure('Units','Normalized','Position',[0 0 1 1])
        text(-1.5,1.1,'SITES OF ELEMENTARY IDEAS ON...
 THE CIRCUMFERENCE','Fontsize', 25,'Color','r')
        hold on
        for g1=1:5:r-1
            for g2=g1+1:5:r
              if (A(g1,g2)>.5)
                   plot([xs(g1),xs(g2)],[ys(g1),ys(g2)])
                   axis off
                   axis equal
                   hold on
              end
             end
             %find ''self''
        end
        for g=1:r
            if strcmp(G(g).name,'self')
                sel=g;
        end
        end
        hold on
        plot(xs(g),ys(g),'*r')
```

Case 8 lets the user display the configuration diagrams of created ideas:

```
case 8
     load c:\mind_data
   figure('Units','Normalized','Position',[0 0 1 1])
   axis off
   clf
   text(.1,.9,'NUMBRER OF CREATED IDEAS :','FontSize',26)
   siz=size(CREATION);
   axis off
   text(.1, .8,num2str(siz(2)),'FontSize',26)
   %text(.1,.6,'Select <idea> number','FontSize',26)
   axis off
  hold on
   number=inputdlg('Enter <idea> number ')
```

```
      number=str2double(number)
      hold off
      content2=CREATION{1,number,1};connector2=CREATION{1,number,2};
      %see_mind_new(content2,connector2,number)
      hold on
      idea_content=CREATION{1,number,1};
idea_connector=CREATION{1,number,2};
          see_mind_mod(idea_content,idea_connector)
          N=radix2num(idea_content(:,2),r)
          text(.1,.7,['IDEA WITH GOEDEL NUMBER ',num2str(N)],...
'FontSize',30,'Color','b')
          pause
end
```

The DEVELOP option takes a long time to execute.

```
 case 9
        load('C:\mind_data');
        A_old=A;
        close all
     clf
     duration=menu(['HOW MANY HOURS OF DEVELOPMENT  ? '],...
'1','2','3','4');
     duration=duration*12;%change 12 to 3600
     t0=clock;genre_old=1;
      while etime(clock,t0)<duration
            genre=select(ones(1,9)./9);

            content=[];
            %create thought germ ''content,connector''
            [content,connector]=think2(genre);
            [content,connector]=add_generator_up_Q(content,...
connector,genre);
        [content,connector]=add_generator_up_Q(content,...
connector,genre);
            w=[];
            if isempty(content)

            else
            v=content(:,2);n_v=length(v);
            k=1:n_v;
            g=G(v(k));
            w=[w,g.level];
            if all(ismember(w,1))

            end
            end
            %is any down bond open?
            found=1;
            while found==1
```

```
                    [i,h,omega,found]=find_open_down_bond(content,...
connector);
                    if found==0
                     else
                             [content,connector,found]=connect_down_bond...
(content,connector, i,h,omega,Q_theme);
                    end

             end
               close
        [content,connector]=add_generator_up_Q(content,...
connector,genre);
        [content,connector]=add_generator_up_Q(content,...
connector,genre);
            genre_old=genre;
         end

    clear content connector omega genre theme
    clear content1 connector1
    save c:\mind_data
    A_new=A;
     angles=2*pi.*[0:r-1]./r;
        xs=cos(angles);ys=sin(angles);
         figure('Units','Normalized','Position',[0 0 1 1])
    subplot(1,2,1),text(-1.5,1.1,'BEFORE...
','Fontsize', 25,'Color','r')


         hold on
         for g1=1:5:r-1
            for g2=g1+1:5:r
             if (A_old(g1,g2)>.5)
                 plot([xs(g1),xs(g2)],[ys(g1),ys(g2)])
                 axis off
                 axis equal
                 hold on
              end
             end
             %find ''self''
         end
         for g=1:r
             if strcmp(G(g).name,'self')
                 sel=g;
         end
         end
         hold on
         plot(xs(g),ys(g),'*r')

         subplot(1,2,2),text(-1.5,1.1,'...AND AFTER',....
'Fontsize', 25,'Color','r')
```

```
%figure('Units','Normalized','Position',[0 0 1 1])

hold on
for g1=1:5:r-1
  for g2=g1+1:5:r
    if (A_new(g1,g2)>.5)
        plot([xs(g1),xs(g2)],[ys(g1),ys(g2)])
        axis off
        axis equal
        hold on
     end
    end
    %find ''self''
  end
  for g=1:r
    if strcmp(G(g).name,'self')
        sel=g;
  end
  end
  hold on
  plot(xs(g),ys(g),'*r')

  pause
```

```
close all
```

The primary function "think" calls a secondary function "think1" that grows a mind germ and then applies the COMPLETION operation to it:

```
function [content,connector]=think1
%simulates GOLEM for given theme of thoughts
content=[];load c:\mind_data
%create thought germ ''content,connector''
[content,connector,Q_theme]=build_thought;
see_mind_germ(content,[])
pause(3)
close all
w=[];
v=content(:,2);n_v=length(v);
k=1:n_v
g=G(v(k));
w=[w,g.level];
ismember(w,1);
if all(ismember(w,1))
    figure('Units','Normalized','Position',[0 0 1 1])
    text(.2,.1 ,['STOP THINKING!   NO OPEN BONDS!'],...
'Color','r','FontSize',20)
    axis off
```

199

```
      pause(1)
      return
end


%is any down bond open?
found=1;
while found==1
    [i,h,omega,found]=find_open_down_bond(content,connector);%_mod?

    if found==0
        'not found'
            see_mind(content,connector)

            pause(1)
            close all
            return
    elseif found==1
        'found'
        [content,connector,found]=connect_down_bond(content,...
connector, i,h,omega,Q_theme);
        see_mind(content,connector)
        pause(1)
    end

end
```

....

E.2   Simple Moves

Among the simple moves is adding a connector:

```
function [content,connector]=add_connector_new(content,connector)
%differs from ''add_g'' in that conntent is not changed
load('C:\mind_data');
if isempty(content)
  return
  else
n=length(content(:,1));
for i1=1:n
   for i2=1:n
   if isempty(connector)
      connector=[1,1,1];%this cludge to avoid error
   else
       h1=content(i1,1);h2=content(i2,1);...
g1=content(i1,2);g2=content(i2,2);
       level1=G(g1).level;level2=G(g2).level;
       if level1==level2+1
          for j=1:3
              is_old=any((connector(:,1)==h1)&(connector(:,2)==h2));
              is_old=is_old|any((connector(:,1)==h1)&...
(connector(:,3)==j));
              reg=connection_regular_new(i1,i2,j,content,...
```

```
connector,g_mod,mod_transfer);
            answer=(~is_old)&(g1~=g2)&(h1~=h2)&reg;
                if answer
                    connector=[connector;[h1,h2,j]];
                end
        end
        end

    end
end
end

end
```

Similarly the functions *add_generator_down* and *add_generator_down_Q* add new generators downwards. The qualifier "*Q*" here indicates that the theme driven "*Q*" vector should be used.

```
function [content,connector]=add_generator_down_Q(content,...
connector,theme)
%executes theme driven associations, downwards ideas
%NOTE: ''connection_regular_new'' has not yet been included
load('C:\mind_data');
gs=set_gs_in_mods(theme,gs_in_mod);

Q(gs)=20;
if isempty(content)
    Q=Q./sum(Q);g=select(Q);
    content=[1,g];
    return
else
    %select one of the gens in ''content''
    n=length(content(:,1));i=select(ones(1,n)./n);
    g=content(i,2);h=content(i,1);
    mod=g_mod(g);
    to_g_downs=[gs_in_mod{mod_transfer{mod,1}},...
gs_in_mod{mod_transfer{mod,2}},...
        gs_in_mod{mod_transfer{mod,3}}];

    %now try to connect down to each of these gens
    probs=[];
    if isempty(to_g_downs)
        return
    else
    end

    n_to_g_downs=length(to_g_downs);
    for nu=1:n_to_g_downs
        prob=Q(to_g_downs(nu))*mu/(n+1);
prob= prob*A(g,to_g_downs(nu))^(1/T);probs=[probs,prob];
```

```
    end
    probs=[probs,1];
    probs=probs./sum(probs);
    nu=select(probs);
    if nu==n_to_g_downs+1
        return
        end
        g_to=to_g_downs(nu);
        new_h=max(content(:,1))+1;

 content=[content;[new_h,g_to]];

mod1=g_mod(g_to);
if ~isempty(connector)
    for j=1:3
        is_old=any((connector(:,1)==h)&(connector(:,2)==new_h));
            is_old=is_old|any((connector(:,1)==h)&...
(connector(:,3)==j));
    if (~is_old)&ismember(mod1,mod_transfer{mod,j});
        connector=[connector;[h,new_h,j]];

    else
    end
      end
else

end
end

function [content,connector]=add_generator_up_Q(content,...
connector,theme)
%executes theme driven thinking upwards ideas
load('C:\mind_data');
gs=set_gs_in_mods(theme,gs_in_mod);
Q(gs)=20;
if isempty(content)
    Q=Q./sum(Q);g=select(Q);
    content=[1,g];
else
    %select one of the gens in ''content''
    n=length(content(:,1));i=select([1:n]./n);
h=content(i,1);g=content(i,2);...
        mod=g_mod(g);
    mod_ups=mod_transfer_inv{mod};
    n_mod_ups=length(mod_ups);
    to_g_ups=[];
    %find generators up from which connection may be created
    for m=1:n_mod_ups
        to_g_ups=[to_g_ups,gs_in_mod{mod_ups(m)}];
    end
```

```
    %now try to connect up to each of these gens
    n_to_g_ups=length(to_g_ups);
    probs=[];
    if isempty(to_g_ups)
        return
    else
    end

    for nu=1:n_to_g_ups
        prob=Q(to_g_ups(nu))*mu/(n+1);
prob= prob*A(g,to_g_ups(nu))^(1/T);
probs=[probs,prob];
    end
    probs=probs./sum(probs);probs=[probs,1];
    nu=select(probs);
    if nu==n_to_g_ups+1
        return
    end
    new_h=max(content(:,1))+1;
        g_to=to_g_ups(nu);
        mod1=g_mod(g_to);
            for j=1:3
                h=content(i,1);
                if isempty(connector)
                    connector=[connector;[new_h,h,j]]
                else
                is_old=any((connector(:,1)==new_h)&(connector(:,2)==h));
                is_old=is_old|any((connector(:,1)==new_h)&...
(connector(:,3)==j));
                if (~is_old)&ismember(mod,mod_transfer{mod1,j});
                        connector=[connector;[new_h,h,j]];
                    end
                end
            end
    content=[content;[new_h,g_to]];
end
```

A thought germ is created by "build_thought":

```
function [content,connector,Q_theme]=build_thought
% computes new thought from scratch (enpty ''content'')...
% according to PRINIPLES
%executes theme driven associations
%NOTE: ''connection_regular_new'' has not yet been included
load C:\mind_data ;

%find gnerators in various levels
[L1,L2,L3,L4]=get_levels(G);

%select theme
```

```
number=menu('Select Theme of Mind','To Have and Have Not',...
'Love and Hate',...
        'Sport','Business','Study','Health','Pets',...
'Conversation','Politics');
    theme=THEMES{1,number,:};
%find generators in ''theme''
gs=set_gs_in_mods(theme,gs_in_mod);content=[];connector=[];
Q(gs)=20;Q_theme=Q;

%thinking power defined in terms of size of ''thought_germ''
prob_germ1=1./[1:4];prob_germ1=prob_germ1./sum(prob_germ1);
n_germ1=select(prob_germ1);


  %form sample of size ''n_germ'' on level 1
  level = 1;
  gs1=intersect(gs,L1);
  sample1=[];Q1=Q(gs1);sampl1=[];
  if ~isempty(gs1)
  for k=1:n_germ1
      sample1=[sample1,select(Q1./sum(Q1))];
  end
  sampl1=gs1(sample1);
end

  %now level 2
  prob_germ2=1./[1:4];prob_germ2=prob_germ2./sum(prob_germ2);
  n_germ2=select(prob_germ2)-1;
  gs2=intersect(gs,L2);
  sample2=[];Q2=Q(gs2);sapl2=[];
  if ~isempty(gs2)
  for k=1:n_germ2
      sample2=[sample2,select(Q2./sum(Q2))];
  end
  sampl2=gs2(sample2);
end

  %now level 3
  prob_germ3=3./[1:2];prob_germ3=prob_germ3./sum(prob_germ3);
  n_germ3=select(prob_germ3)-1;
  gs3=intersect(gs,L3);
  sample3=[];Q3=Q(gs3);sampl3=[];
  if ~isempty(gs3)
  for k=1:n_germ3
      sample3=[sample3,select(Q3./sum(Q3))];
  end
  sampl3=gs3(sample3);
end

  %now level 4
```

```
prob_germ4=1./[1:1];prob_germ4=prob_germ4./sum(prob_germ4);
n_germ4=select(prob_germ4)-1;
gs4=intersect(gs,L4);
sample4=[];Q4=Q(gs4);sampl4=[];
if ~isempty(gs4)
for k=1:n_germ4
    sample4=[sample4,select(Q4./sum(Q4))];
end
sampl4=gs4(sample4);
end


n=length(sampl1)+length(sampl2)+length(sampl3)+length(sampl4);
content(:,1)=[1:n]';
if ~isempty(content)
content(:,2)=[sampl1,sampl2,sampl3,sampl4]'
end
```

An auxiliary program finds connected components in configuration; code from www.math.wsu.edu/faculty/tsat/matlab.html.

```
function [c,v] = conn_comp(a,tol)
warning off
%        Finds the strongly connected sets of vertices
%                in the DI-rected G-raph of A
%        c = 0-1 matrix displaying accessibility
%        v = displays the equivalent classes

%make symmetric
a=(a+a')/2;

[m,n] = size(a);
if m~=n 'Not a Square Matrix', return, end
b=abs(a); o=ones(size(a)); x=zeros(1,n);
%msg='The Matrix is Irreducible !';
%v='Connected Directed Graph !';
v=zeros(1,m);v(1,:)=1:m;
if (nargin==1) tol=n*eps*norm(a,'inf'); end

% Create a companion matrix
b>tol*o; c=ans; if (c==o)  return, end
% Compute accessibility in at most n-step paths
for k=1:n
    for j=1:n
        for i=1:n
            % If index i accesses j, where can you go ?
            if c(i,j) > 0  c(i,:) = c(i,:)+c(j,:); end
        end
    end
```

```
end
% Create a 0-1 matrix with the above information
c>zeros(size(a)); c=ans; if (c==o) return, end

% Identify equivalence classes
d=c.*c'+eye(size(a)); d>zeros(size(a)); d=ans;
v=zeros(size(a));
for i=1:n find(d(i,:)); ans(n)=0; v(i,:)=ans; end

% Eliminate displaying of identical rows
i=1;
while(i<n)
      for k=i+1:n
            if v(k,1) == v(i,1)
                  v(k,:)=x;
            end
      end
      i=i+1;
end
j=1;
for i=1:n
      if v(i,1)>0
         h(j,:)=v(i,:);
         j=j+1;
      end
end
v=h;
%end
```

To connect bonds down:

```
function [content,connector,found]=connect_down_bond(content,...
connector, i,h,omega,Q_theme)
%finds generator to connect to open down bond (i,h,omega)
load c:\mind_data G mod_transfer gs_in_mod Q A T
g=content(i,2);n=length(content(:,1));
if ~isempty(connector)
    m=length(connector(:,1));
else m=0;
end


%connect generator to what? Set of ''to_gs'' =v;
s=G(g);
mod=s.modality;
to_mods=mod_transfer{mod,omega};to_gs=gs_in_mod(to_mods);...
 n_to_gs=length(to_gs);
%connect to g's?
v=[];
for nu=1:n_to_gs
    v=[v,to_gs{nu}];
end
```

```
to_gs=v;
old_gs= ismember(content(:,2),to_gs);
    if any(old_gs)
        u=content(:,1);v=content(:,2);
        to_h=u(logical(old_gs));
        to_g=v(logical(old_gs));n_to_h=length(to_h)

        %random selection
        probs=[];
        for nu=1:n_to_h
    prob=Q(v(nu))*n/(n+1);
prob= prob*A(g,v(nu))^(1/T);
probs=[probs,prob];
        end
    probs=probs./sum(probs);
    nu=select(probs);

        to_h=to_h(nu);
        t=isempty(connector);
        if t==1
            connector=[h,to_h,omega];
            found=1;
            return
        end
        already_connected=(connector(:,1)==
        h)&(connector(:,2)==to_h);%error?
        if ~any(already_connected)
                connector=[connector;[h,to_h,omega]];
                found=1;
                return
         end
        %else find new g to connect to
    end
        %sample from probs over set ''to_gs''
        probs=[];
        for mu=1:n_to_gs
            prob=Q_theme(to_gs(mu))*mu/(n+1);prob=...
 prob*A(g,to_gs(mu))^(1/T);probs=[probs,prob];
            probs=[probs,prob];
        end
        probs=probs./sum(probs);
        new_g=select(probs);new_g=to_gs(new_g);
        %connect this ''new_g'' to old content, connector
        content=[content;[max(content(:,1))+1,new_g]];r=1:3;

            connector=[connector;[h,max(content(:,1)),omega]];...
%note that ''content''already been incremented
            found=1;
```

To verifiy that down connection is regular:

```
function answer=connection_regular_new(i1,i2,j,content,...
connector,g_mod,mod_transfer)
%finds whether proposed connection i1->i2  for ''j''th ...
down bond is regular
answer=0;
if i1==i2
    %same generator?
    return
end

%first check whether modalities satisfy regularity
h1=content(i1,1);h2=content(i2,2);
g1=content(i1,2);g2=content(i2,2);
mod1=g_mod(g1);mod2=g_mod(g2);
mod=mod_transfer{mod1,j};
if ismember(mod2,mod)
                answer=1;
                return
end
```

To create new idea:

```
    function class_idea = create_idea
    %Use local coordinates for idea. Only 2_top_idea allowed
    omega=input(' Down arity = \n');idea_class=cell(1,omega);
    load('C:\mind_data');
    r=length(G);Q=ones(1,r);
    for l=1:omega+1
        svar= input(['for bond no. ', num2str(l),' modality (1)...
 or generators (2) ? \n'])
    if svar==1
        mod=input('modality = ? \n');
        idea_class{1,l}=gs_in_mod(mod)
    elseif svar ==2
        gs=input('give vector of generators \n')
        idea_class{1,l}=gs;
    end
end
class_idea=idea_class;
```

To delete generator from *G*, use with caution:

```
function delete_g(g,G)
%deletes single generator ''g'' in ''G''
r=length(G);
v=[[1:g-1],[g+1:r]];
G(v);
```

To delete generator with its connections:

```
function [content,connector]=delete_generator_connections(content,...
connector)
%this program deletes generator and associated connections
load('c:\mind_data');
if isempty(content)
   return
   else
      n=length(content(:,1));
      %select generator

      i_del=select(ones(1,n)./(n));%in i-coordiantes
      g=content(i_del,2);
if i_del>=n
   return
end

if isempty(connector)
   prob_del=(n/mu)/Q(g);  %check this!
   prob_del=prob_del/(1+prob_del);
   if select([prob_del,1-prob_del])
      content=content([1:i_del-1,i_del+1],:);
      return
   end
else
   m=length(connector(:,1));
   %bonds down to this generator from others above
   h=content(i_del,1);
j_above=find(connector(:,2)==h);%in j-coordinates
l_above=length(j_above);
product=n/(mu*Q(g));
for j=1:l_above
   j=j_above(j);h1=connector(j,1);
   i1=find(content(:,1)==h1);i2=find(content(:,1)==h);
   g1=content(i1,2);g2=content(i2,2);
   product=product*(A(g1,g2))^(-1/T);
end

%bonds up to this generator from others below
j_down=find(connector(:,1)==h);%in j-doordinates
l_down=length(j_down);
for j=1:l_down
   j=j_down(j);h2=connector(j,2);
   i1=find(content(:,1)==h);i2=find(content(:,1)==h2);
   g1=content(i1,2);g2=content(i2,2);
   product=product*(A(g1,g2))^(-1/T);
end

prob_del=product;%check this!
prob_del=prob_del/(1+prob_del);
```

```
answer=select([prob_del,1-prob_del]);
if answer==1
   content=content([1:i_del-1,i_del+1:n],:);
   connector=connector(setdiff([1:m],[j_above',j_down']),:);

else
end
end
end
```

To delete generators but keep external inputs:

```
function [content,connector]=delete_generator_keep_input(content,...
connector)
%this program has been written so that a ...
%simple modification (defining ''n_input)
% will make the inputted ''content'' stay unchanged
load c:\matlabr12\golem2\mind_data2 A G Q T g_mod mod_transfer mu;
if isempty(content)
   return
   else
      n=length(content(:,1));
      %select generator, not input
n_input=0;
    i_del=n_input+select(ones(1,n-n_input)./(n-n_input));
%in i-coordiantes
      g=content(i_del,2);
if i_del>n
   return
end

if isempty(connector)
   prob_del=(n/mu)/Q(g);%check this!
   prob_del=prob_del/(1+prob_del);
   if select([prob_del,1-prob_del])
      content=content([1:i_del-1,i_del+1],:);
      return
   end
else
   m=length(connector(:,1));

   %bonds down to this generator from others above
   h=content(i_del,1);
j_above=find(connector(:,2)==h);%in j-coordinates
l_above=length(j_above);
product=n/(mu*Q(g));
for j=1:l_above
   j=j_above(j);h1=connector(j,1);
   i1=find(content(:,1)==h1);i2=find(content(:,1)==h);
   g1=content(i1,2);g2=content(i2,2);
   product=product*(A(g1,g2))^(-1/T);
```

```
end

%bonds up to this generator from others below
j_down=find(connector(:,1)==h);%in j-doordinates
l_down=length(j_down);
for j=1:l_down
    j=j_down(j);h2=connector(j,2);
    i1=find(content(:,1)==h);i2=find(content(:,1)==h2);
    g1=content(i1,2);g2=content(i2,2);
    product=product*(A(g1,g2))^(-1/T);
end

prob_del=product;%check this!
prob_del=prob_del/(1+prob_del)
answer=select([prob_del,1-prob_del]);
if answer==1
    content=content([1:i_del-1,i_del+1:n],:);
    connector=connector(setdiff([1:m],[j_above',j_down']),:);

else
end
end
end
```

To find idea in "thought":

```
function [idea_content,idea_connector]=...
get_idea_thought(content,connector)
%displays one of the ''ideas'' in ''thought''
[top_2ideas_g,top_2ideas_h]=get_top_2ideas(content,connector);
[idea_content,idea_connector]=single_idea(content,...
connector,top_2ideas_g{1},top_2ideas_h{1});
```

To find dominating thought:

```
function [content1,connector1]=dom_thought(content,connector)
%computes connected components in thought chatter and finds largest
%component
if isempty(connector) | isempty(content)
    content1=[];connector1=[];
    return
else
end

n=length(content(:,1));m=length(connector(:,1));
%create DI-graph
graph=zeros(n);
for j=1:m
    h1=connector(j,1);h2=connector(j,2);
    i1=find(content(:,1)==h1);
```

```
        i2=find(content(:,1)==h2);
        graph(i1,i2)=1;
end

%find connected components
[c,v]=conn_comp(graph);
    ls=sum((v>0),2);
    [y,i]=max(ls);
    is=v(i,:);is=find(is);is=v(i,is);
    if ischar(is)
        content1=content;connector1=connector;
        return
    else
    end
 content1=content(is,:);
 %find rows in new connector1
 connector1=[];
 for j=1:m
     if ismember(connector(j,1),content1(:,1))&...
ismember(connector(j,2),content(:,1))
         connector1=[connector1;connector(j,:)];
     end
 end
```

To get template for driver:

```
function [content,connector]=driver_template(driver,content,...
connector,content_idea,connector_idea )
%transforms mental state with driver expressed as...
 ''content_idea''+''connector_idea''
%into new mental state.
% use ''name'' instead of ''driver'' in line 0 (as character string)
load(['\matlabr12\golem2\',driver])
s=select([activation_probability,1-activation_probability]);
if s==2
    return
end
load \matlabr12\golem2\mind_data2 class_idea
%check if driver is applicable to this drive
x=size(class_idea)

omega_driver=x(1);applicable=1;
for k=1:omega_driver
    if ~ismember(content_idea(k,:),class_idea(k,:))
%perhaps cell structures?
        applicable=0;
    end

    if applicable
```

```
r=length(G);n=length(content(:,1));m=length(connector(:,1));
%only adds new connections inside idea; use i_ and j_coordinates
%formats:change_idea cell array (2,n_idea) with values in first row
% 'delete' meaning delete this generator
%'same' meaning same generator, unchanged
%'replace' by g
%'random' set of g's, randomly select one from this set
%in second row column 3 g-value;
% in second row column 4 set of g'values, other columns []
%format of ad_content: 2-column matrix ,...
% first column max(content(:,1))+1,
%second column g-values
%format of ad_connector: 3-column martix with i-coordinates...
% in first two columns, bond coordinate in third column
%format delet_connector: vector of j-coordinates

%keep configuration minus ''idea''
keep_h=setdiff(content(:,1),content_idea(:,1));
keep_i=find(ismember(content(:,1),keep_h));
keep_content=content(keep_i,:);
keep_connector=find(ismember(connector(:,1),keep_h)&...
ismember(connector(:,2),keep_h));
keep_connector=connector(keep_connector,:);
between1=ismember(connector(:,1),keep_h)&ismember(connector(:,2),...
content_idea(:,1));
between2=ismember(connector(:,2),keep_h)&ismember(connector(:,1),...
content_idea(:,1));
keep_idea_connector=connector(find(between1'|between2'),:);
n=length(content(:,1));m=length(connector(:,1));
n_idea=length(content_idea(:,1));
n_ad=length(ad_content);
m_idea=length(connector_idea);
m_add=length(ad_connector);
m_delet=length(delet_connector);

%begin by changing values (no deletion yet)
del=zeros(1,n);
for i=1:n_idea
    if strcmp(change_idea{i,1},'delete')
        del(i)=1;
    elseif strcmp(change_idea{i,1},'same');
    elseif strcmp(change_idea{i,1},'replace')
        content_idea(i,2)=change_idea{i,2};
    elseif strcmp(change_idea{i,1},'random')
        new_set=change_idea{i,4};n_new_set=length(new_set);
        choose=select([1:n_new_set]./n_new_set);
        content_idea(i,2)=new_set(choose);
    end

end
```

```
%then add new generators
content_idea=[content_idea;ad_content];
%then add new connections
if m_add>0
for j=1:m_add
    h1=ad_connector(j,1); h2=ad_connector(j,2);
    %h1=content_idea(1,i1); h2=content(1,i2);
    connector_idea=[connector_idea;[h1,h2,b]];
    end

end

v=setdiff([1:n_idea],del);
content_idea = content_idea(v,:);

%put transformed ''idea'' back into configuration
new_content=[keep_content;content_idea];

new_connector=keep_connector;
if ~isempty(connector_idea)
new_connector=[keep_connector;connector_idea];
end
if ~isempty(keep_idea_connector)
new_connector=[new_connector;keep_idea_connector];
end
end
end
content=new_content;
connector=new_connector;
```

Executes driver:

```
function [content,connector]=execute_driver(driver,...
content,connector)
%executes driver named ''driver'' for (total) idea
load('c:\mind_data')
if isempty(connector)
    return
end
n=length(content(:,1));m=length(connector(:,1));
[top_2ideas_g,top_2ideas_h]=get_top_2ideas(content,connector);
 %these are the top_2ideas
    n_ideas=length(top_2ideas_g);
belongs_to_domain=zeros(1,n_ideas);
    domain=driver{6};
    %find if any of the top_2ideas in idea...
% belongs to ''domain'' of ''driver''
    %check each entry in of top_2idea w.r.t. ''domain'' of driver
    for k=1:n_ideas
```

```
        gs=top_2ideas_g{1,k,:}; n_gs=length(gs);above=gs(1);
        below=[];hs=top_2ideas_h{1,k,:};
        driv=driver{1};
        belongs_to_domain(k)= ismember(above,domain{1});
        for n=2:n_gs
            belongs_to_domain(k)=belongs_to_domain(k)&...
(ismember(gs(k),domain{k}))|isempty(domain{k});
        end
        %belongs_to_domain
        if ~belongs_to_domain
            return
        end
        first_idea=min(find(belongs_to_domain));
        gs=top_2ideas_g{1,first_idea,:};
hs=top_2ideas_h{1,first_idea,:};n_idea=length(gs);
        %do not execute ''driver'' for the first idea with probability
        if rand(1)>driver{5}
            return
        end
    end


        %now execute ''change_idea'' of ''driver''
        change_idea=driver{1};dels=[];%i-numbers of deletions
        for i=1:n_idea %enumerates generators in sub-idea
            if strcmp(change_idea{i,1},'delete')
                dels(i)=1;
            else if strcmp(change_idea{i,1},'same')
            elseif  strcmp(change_idea{i,1},'replace')
                i_value=  find(content(:,1)== hs(i));
g_new=change_idea{i,2};
                content(i_value,2)=g_new
            elseif strcmp(change_idea{i,1},'random')
                i_value=  find(content(:,1)== hs(i));
                g_set=change_idea{i,2};g_set_n=length(g_set);
                choose=select([1:g_set_n]./g_set_n);
                g_new=g_set(choose);
                content(i_value,2)=g_new;
            end
        end

        %deletes generators with dels==1 (i-numbers in sub-idea)
        del_h=hs(dels);
        if ~isempty(del_h)
            i_dels=[];
            %delete generators
            for k=1:n
                i_dels=[i_dels,find(content(:,1)==del_h)];
            content=content(setdiff([1:n],i_dels),:);
            end
```

```
            %delete connections
            j_s=[];
            for j=1:m
                j_s=[j_s,find(ismember(connector(j,1),del_h))|...
                        find(ismember(connector(j,2),del_h))];
            end
                connector=connector(setdiff([1:m],j_s),:);
        end

        %add new generators
        ad_content=driver{2};
        content=[content;ad_content]

        %add new connectors
        ad_connector=driver{3};
        connector=[connector;ad_connector];

        %delete connectors in ''idea''
        delet_connector=driver{4};
        j=find((connector(:,1)==hs(1))&(connector(:,3)...
==delet_connector));
        m=length(connector(:,1));
        connector=connector(setdiff([1:m],j),:);

    end
```

To find element in "*G*":

```
function find_g
%searches for generator number with given name
name=input( 'specify name  \n','s')
load c:\mind_data
r=length(G);
for g=1:r
    if strcmp(G(g).name,name)
        g
    end
end
```

To find open bond downwards:

```
function [i,h,omega,found]=find_open_down_bond(content,connector)
%prepares for completing the given thought...
 expressed as content,connectorn
%by searching for open down bond
if isempty(content)
    i=1;h=1;omega=1;found=0;not_found=1;
    'EMPTY THOUGHT'
    return
end
```

```
%find''down'' open down-bonds
load c:\mind_data
n=length(content(:,1));found=0;

for i=1:n
    h=content(i,1);g=content(i,2);mod=g_mod(g);
    arity=mod_omegas(mod);
    if (arity >0) & (~isempty(connector))
        m=length(connector(:,1));
        for omega=1:arity
            v=(connector(:,1)==h)&(connector(:,3)==omega);
            if all(v==0)
                found=1;
                return
            end
        end
    end
end
if isempty(connector)
    for i=1:n
        h=content(i,1);g=content(i,2);mod=g_mod(g);
        arity=mod_omegas(mod);
        if arity>0
          found=1;
          omega=1;
        end
        omega=1;
    end
end
```

Computes level sets in "*G*":

```
function [L1,L2,L3,L4]=get_levels(G);
%computes G-sets for level=1,1...
r=length(G);L1=[];L2=[];L3=[];L4=[];
for g=1:r
   l=G(g).level;
   if l==1
      L1=[L1,g];
   elseif l==2
      L2=[L2,g];
   elseif l==3
      L3=[L3,g];
   elseif l==4
      L4=[L4,g];
   end

end
```

To compute inverse of transformation "mod_transfer":

```
function mod_transfer_inv=get_mod_transfer_inv(mod_transfer)
%computs inverse of ''mod_transfer''
n_mods=length(mod_transfer);mod_transfer_inv=cell(1,n_mods);
 n_mods
for mod=1:n_mods
for k=1:n_mods
   for j=1:3
        if ismember(mod,mod_transfer{k,j})
            mod_transfer_inv{mod}=[ mod_transfer_inv{mod},k];
        else
        end
   end
end
end
```

To find top-ideas in "thought":

```
function [top_2ideas_g,top_2ideas_h]=get_top_2ideas(content,...
connector)
%computes only second level ideas;
%this MIND is intellectually challenged and
%cannot think about abstractions of level greater than two
%produces only complete ideas
if isempty(connector)
    top_2ideas_g=[];top_2ideas_h=[];
    figure('Units','Normalized','Position',[0 0 1 1])
    axis off
    text(.2,.5,'No top-2ideas','FontSize',32)
    pause(2)
    return
end
load('c:\mind_data')
tops_i=find(ismember(content(:,2),L2));%in i-coordinates
tops_g=content(tops_i,2);
%above in g-coordinates
tops_h=content(tops_i,1);
% above is in h-coordinates
n_tops=length(tops_i); top_2ideas_g=cell(1,n_tops);
top_2ideas_h=cell(1,n_tops);
for k=1:n_tops
   top_2ideas_g{1,k,1}=tops_g(k);
   top_2ideas_h{1,k,1}=tops_h(k);
   top_g=tops_g(k);top_h=tops_h(k);
mod=G(top_g).modality;omega=mod_omegas(mod);
   f=find((connector(:,1)==top_h)&(connector(:,3)==1));
   if ~isempty(f)
   f1=connector(f,2);i=find(content(:,1)==f1);f=content(i,2);
   top_2ideas_g{1,k,:}=[top_2ideas_g{1,k,:},f];
```

```
    top_2ideas_h{1,k,:}=[top_2ideas_h{1,k,:},f1];
    end
    f=find((connector(:,1)==top_h)&(connector(:,3)==2));
    if ~isempty(f)
    f1=connector(f,2);i=find(content(:,1)==f1);f=content(i,2);
    top_2ideas_g{1,k,:}=[top_2ideas_g{1,k,:},f];
    top_2ideas_h{1,k,:}=[top_2ideas_h{1,k,:},f1];
    end
    f=find((connector(:,1)==top_h)&(connector(:,3)==3));
    if ~isempty(f)
    f1=connector(f,2);i=find(content(:,1)==f1);f=content(i,2);
    top_2ideas_g{1,k,:}=[top_2ideas_g{1,k,:},f];
    top_2ideas_h{1,k,:}=[top_2ideas_h{1,k,:},f1];
    end
end

%find complete ideas
complete=zeros(1,n_tops);
for k=1:n_tops
    v=top_2ideas_g{1,k,:};
    top=v(1);mod=g_mod(top);omega=mod_omegas(mod);
    if (length(v)==1+omega)
        complete(k)=1;
    end
end

%now keep only complete ideas
top_2ideas_g=top_2ideas_g(find(complete));
top_2ideas_h=top_2ideas_h(find(complete));
```

....

To compute the energy function we execute:

```
function E=energy(content,connector)
%computes energy in thought
load('c:\mind_data')
if isempty(content)
    E=0;
```

Made in the USA
Middletown, DE
05 November 2017